Conference Board of the Mathematical Sciences
REGIONAL CONFERENCE SERIES IN MATHEMATICS

supported by the
National Science Foundation

Number 34

LECTURE NOTES ON NIL-THETA FUNCTIONS

by

Louis Auslander

Published for the
Conference Board of the Mathematical Sciences
by the
American Mathematical Society
Providence, Rhode Island

Expository Lectures
from the CBMS Regional Conference
held at Cleveland State University
June 5–10, 1977

AMS (MOS) subject classifications (1970). Primary 14–XX, 22–XX, 33–XX, 42–XX

Library of Congress Cataloging in Publication Data

CIP

Auslander, Louis.
Lecture notes on nil-theta functions.

(Regional conference series in mathematics ; no. 34)
"Expository lectures from the CBMS regional conference held at Cleveland State University, June 5-10, 1977."
Bibliography: p.
1. Functions, Theta. 2. Lie groups. 3. Fourier analysis. I. Title. II. Series.
QA1.R33 no. 34 [QA345] 510'.8s [515'.984] 77-16471
ISBN 0-8218-1684-5

Printed in the United States of America

Other Monographs in this Series

No. 1. Irving Kaplansky: *Algebraic and analytic aspects of operator algebras*
2. Gilbert Baumslag: *Lecture notes on nilpotent groups*
3. Lawrence Markus: *Lectures in differentiable dynamics*
4. H. S. M. Coxeter: *Twisted honeycombs*
5. George W. Whitehead: *Recent advances in homotopy theory*
6. Walter Rudin: *Lectures on the edge-of-the-wedge theorem*
7. Yozo Matsushima: *Holomorphic vector fields on compact Kähler manifolds*
8. Peter Hilton: *Lectures in homological algebra*
9. I. N. Herstein: *Notes from a ring theory conference*
10. Branko Grünbaum: *Arrangements and spreads*
11. Irving Glicksberg: *Recent results on function algebras*
12. Barbara L. Osofsky: *Homological dimensions of modules*
13. Michael Rabin: *Automata on infinite objects and Church's problem*
14. Sigurdur Helgason: *Analysis on Lie groups and homogeneous spaces*
15. R. G. Douglas: *Banach algebra techniques in the theory of Toeplitz operators*
16. Joseph L. Taylor: *Measure algebras*
17. Louis Nirenberg: *Lectures on linear partial differential equations*
18. Avner Friedman: *Differential games*
19. Béla Sz.-Nagy: *Unitary dilations of Hilbert space operators and related topics*
20. Hyman Bass: *Introduction to some methods of algebraic K-theory*
21. Wilhelm Stoll: *Holomorphic functions of finite order in several complex variables*
22. O. T. O'Meara: *Lectures on linear groups*
23. Mary Ellen Rudin: *Lectures on set theoretic topology*
24. Melvin Hochster: *Topics in the homological theory of modules over commutative rings*
25. Karl W. Gruenberg: *Relation modules of finite groups*
26. Irving Reiner: *Class groups and Picard groups of group rings and orders*
27. H. Blaine Lawson, Jr.: *The quantitative theory of foliations*
28. T. A. Chapman: *Lectures on Hilbert cube manifolds*
29. Alan Weinstein: *Lectures on symplectic manifolds*
30. Aleksander Pelczynski: *Banach spaces of analytic functions and absolutely summing operators*
31. Ronald R. Coifman and Guido Weiss: *Transference methods in analysis*
32. Wolfgang M. Schmidt: *Small fractional parts of polynomials*
33. George Glauberman: *Factorizations in local subgroups of finite groups*
34. Louis Auslander: *Lecture notes on nil-theta functions*

Math Sep

TABLE OF CONTENTS

Chapters

I. Foundations. 1

1. Bilinear Forms and Presentations of Certain 2-Step Nilpotent Lie Groups. 1
2. Discrete Subgroups of the Heisenberg Group. 6
3. The Automorphism Group of the Heisenberg Group. 14
4. Fundamental Unitary Representations of the Heisenberg Group. 17
5. The Fourier Transform and the Weil-Brezin Map. 22
6. Distinguished Subspaces and Left Action. 34

II. Jacobi Theta Functions and the Finite Fourier Transform. 48

1. Nil-Theta Functions and Jacobi-Theta Functions. 48
2. The Algebra of the Finite Fourier Transform. 60

III. Abelian Varieties, Nil-Theta and Theta Functions. 69

1. A General Construction and Algebraic Foundations. 69
2. Nil-Theta Functions Associated with a Positive Definite H-morphism of an Abelian Variety. 83
3. The Relation Between Nil-Theta and Classical Theta Functions. 89

Bibliography 95

Introduction

The material in these notes represents joint work with Richard Tolimieri and was much influenced by my previous joint work with Jonathan Brezin. In addition, some of the concepts were developed in a joint effort of myself, Tolimieri and Barry Kolb that was announced in [7]. In presenting the material, I have tried to lay a careful foundation, and I have stressed low-dimensional examples and special computations even when I later prove general results by general techniques. Also, the last two sections are minimally developed with the interested reader being urged to consult Tolimieri [11] for a complete treatment.

The relation of these Notes to the classical literature should be self-evident and the results in [3] will make this more specific for the interested reader. However, these notes do something that may not be so evident. In three important works [14], [12], [13] A. Weil presents a proof of the Plancherel Theorem, a new treatment of Abelian varieties and what we now call the Weil-Brezin map. That all these are inter-related is by no means apparent. In these notes they all become united in the study of nil-theta functions.

Acknowledgement

I would like to thank Cleveland State University for acting as host Institute and Professor Allan Silberger for his efforts that made the Regional Meeting such a success.

I. Foundations.

I.1. Bilinear Forms and Presentations of Certain Two-Step Nilpotent Lie Groups.

Let V^n or V be an n-dimensional real vector space, and let $B : V \times V \to \mathbb{R}$ be a bilinear mapping. We begin by using B to define a nilpotent Lie group which we will denote by $N(B)$. Those people who are familiar with such things will be able to find nilpotent associative algebras lurking in the shadows.

Let $v_i \in V$ and $t_i \in \mathbb{R}$, $i = 1, 2$; we define a multiplication on the set $N(B) = V \times \mathbb{R}$ by the formula

1. $$(v_1, t_1)(v_2, t_2) = (v_1+v_2, t_1+t_2+B(v_1, v_2)).$$

It is straightforward to verify that $N(B)$ is a group under this law of composition with identity $(\underline{0}, 0) \in V \times \mathbb{R}$ and $(v,t)^{-1} = (-v, B(v,v)-t)$.

If $(x_1, \ldots, x_n) = v$ and $B = (a_{ij})$, $i, j = 1, \ldots, n$, relative to this basis of V, then it is easy to verify that

$$((x_1, \ldots, x_n), t) \to \begin{pmatrix} 1 & \Sigma a_{j1}x_j & & \cdots & \Sigma a_{jn}x_j & t \\ & 1 & 0 & \cdots & 0 & x_1 \\ & & 1 & & 0 & \vdots \\ & & & & 1 & x_n \\ & & & & & 1 \end{pmatrix}$$

is a faithful representation of $N(B)$. Clearly the subgroup $Z = \{(0, t) \mid t \in \mathbb{R}\}$ is a normal subgroup of $N(B)$ isomorphic to $\mathbb{R}$ and $N(B)/Z$ is isomorphic to V. Hence, $Z \supset [N(B), N(B)]$, where $[\ ,\ ]$ denotes the commutator subgroup of the group in the bracket. Hence $N(B)$ is a 2-step nilpotent Lie group and if it is not abelian, its commutator subgroup has dimension one.

Let $B = A+S$, where A is the alternating bilinear form $A = \frac{1}{2}(B-B^t)$, where the superscript t denotes the transpose of the form B and S is the symmetric bilinear form $S = \frac{1}{2}(B+B^t)$. We will sometimes also use the notation $B = A(B)+S(B)$.

Now let A be an alternating bilinear form, and let S be a symmetric bilinear form. We may consider $N(A)$ and $N(A+S)$. We will now provide an isomorphism between $N(A)$ and $N(A+S)$. Let a subscript 0 denote that the element is in $N(A)$ and a subscript 1 denote that the element is in $N(A+S)$. Thus

$$(v_1, t_1)_0(v_2, t_2)_0 = (v_1+v_2, t_1+t_2+A(v_1, v_2))_0$$

and

$$(v_1, t_1)_1(v_2, t_2)_1 = (v_1+v_2, t_1+t_2+A(v_1, v_2)+S(v_2, v_2))_1 .$$

Let
$$\pi_S((v, t)_0) = (v, t+\tfrac{1}{2}S(v, v))_1 .$$

Then it is a straightforward exercise to verify that

$$\pi_S : N(A) \to N(A+S)$$

is an isomorphism. An elementary computation gives that

$$\pi_S^{-1} : N(A+S) \to N(A)$$

is given by $\pi_A^{-1}(v, t)_1 = (v, -\frac{1}{2}S(v, v)+t)_0$. Hence $N(A+S_1)$ and $N(A+S_2)$ are isomorphic under the isomorphism $\pi_{S_2} \circ \pi_{S_1}^{-1}((v, t)_1) = (v, \frac{1}{2}S_2(v, v)-\frac{1}{2}S_1(v, v)+t)_2$ where the subscript 2 denotes the group $N(A+S_2)$ and the subscript 1 the group $N(A+S_1)$.

Thus we have reduced the isomorphism problem for the general bilinear form to the case of alternating forms. We may now use the fundamental fact about alternating forms that states that there exists a nonsingular linear transformation

$$C : V \to V$$

such that

$$CAC^t = \begin{pmatrix} 0 & \frac{1}{2}I_r & 0 \\ -\frac{1}{2}I_r & 0 & 0 \\ 0 & 0 & 0 \end{pmatrix}$$

where I_r is the $r \times r$ identity matrix and zeros denote zero matrices of the appropriate sizes. (See [8].) In other words, the kernel of an alternating bilinear form is the only invariant of the form. For most of the rest of these notes, we may assume that all A have been written in the above form. We will now show that the above discussion easily implies that if $z(\ )$ denotes the center of the group in the bracket,

$$\dim z(N(A)) = 1+\dim\{\text{kernel of } A\}.$$

For let $w \in z(N(A))$; then for $(v,0) \in N(A)$

$$(w,0)(v,0) = (w+v, A(w,v))$$

$$(v,0)(w,0) = (w+v, A(v,w)).$$

Hence $A(w,v) = A(v,w) = A(v,w) = 0$ for all $v \in V$ and w is in the kernel of A.

Conversely, let w be in the kernel of A. Since

$$0 = A(w,v) , \qquad v \in V ,$$

We have $(w, 0) \in z(N(A))$.

Hence $\dim z(N(A))$ or the dimension of the kernel of A is the only invariant, and we have established the following:

Theorem I.1.1. $N(B_1)$ and $N(B_2)$ are isomorphic if and only if $A(B_1)$ and $A(B_2)$ have kernels of the same dimension.

Definition. If $\dim V = 2n$ and $A(B)$ is nondegenerate, we will call 2n+1-dimensional group $N(B)$ the Heisenberg group of $\dim 2n+1$.

The Heisenberg group has two presentations that are particularly important in applications. The first is $N(A_0)$ where

$$A_0 = \begin{pmatrix} 0 & \frac{1}{2}I_n \\ -\frac{1}{2}I_n & 0 \end{pmatrix}$$

which we will call the basic presentation.

The second is $N(D) = N(A_0+B_0)$, where

$$B_0 = \begin{pmatrix} 0 & \frac{1}{2}I_n \\ \frac{1}{2}I_n & 0 \end{pmatrix}.$$

In $N(A_0+B_0)$ multiplication is given by

$$(v_1, t_1)(v_2, t_2) = (v_1+v_2, t_1+t_2+v_1 C v_2^t)$$

where $C = \begin{pmatrix} 0 & I_n \\ 0 & 0 \end{pmatrix}$. This will be called dual pairing presentation of the Heisenberg group for reasons that we will make apparent after the following remark. Let V^{2n} have coordinates $(x_1, \ldots, x_n, y_1, \ldots, y_n)$. Then multiplication in $N(D)$ is given by

$$(x_1,\ldots,x_n,y_1,\ldots,y_n,t_1)(a_1,\ldots,a_n,b_1,\ldots,b_n,t_2) = (v_1+v_2, t_1+t_2+\sum_{i=1}^{n} x_i b_i)$$

where $v_1 = (x_1,\ldots,x_n,y_1,\ldots,y_n)$ and $v_2 = (a_1,\ldots,a_n,b_1,\ldots,b_n)$.

We will now discuss our justification for calling $N(D)$ the dual pairing presentation of the Heisenberg group. Let $\mathbb{R}^n$ be the abelian group consisting of n copies of the reals and $\hat{\mathbb{R}}^n$ be its dual group. Let $x \in \mathbb{R}^n$ and $y \in \hat{\mathbb{R}}^n$. Then $D(x,y) = \exp 2\pi i x \cdot y$ is the pairing of $\mathbb{R}^n \times \hat{\mathbb{R}}^n \to \mathbb{T}$, where $\mathbb{T}$ is the circle group, given by the duality of locally compact abelian groups. Let $V = \mathbb{R}^n \oplus \hat{\mathbb{R}}^n$ and define a group G on $V \times \mathbb{T}$ where multiplication is given by

$$((x_1,y_1),\xi_1)((x_2,y_2),\xi_2) = ((x_1+x_2,y_1+y_2),\xi_1 \cdot \xi_2 \cdot e^{2\pi i x_1 \cdot y_2})$$

where $((x,y),\xi) \in \mathbb{R}^n \oplus \hat{\mathbb{R}}^n \times \mathbb{T}$. It is easily verified that the universal covering group of G has the dual pairing presentation of the Heisenberg group.

The basic presentation is intimately related to considering $\mathbb{C}^n = V^{2n}$. Let $(c_1,\ldots,c_n) \in \mathbb{C}^n$. If we define

$$((c_1,\ldots,c_n),t_1)((c_1',\ldots,c_n'),t_2) = ((c_1+c_1',\ldots,c_n+c_n'),t_1+t_2 + \tfrac{1}{2}\mathrm{Im}(c_1\bar{c}_1' + \ldots + c_n\bar{c}_n'))$$

we obtain the basic presentation of the Heisenberg group.

Computations involving the Heisenberg group are usually easiest in its basic presentation. We will list a few facts about the basic presentation.

B.1. The straight lines through the origin in $V^{2n} \times \mathbb{R}$ are the 1-parameter groups in the basic presentation, i.e., the one-parameter groups $g(t)$ in $N(A_0)$ have the form

$$g(t) = \{(vt,ct) \mid v \in V^{2n},\ c \in \mathbb{R} \text{ all } t \in \mathbb{R}\}.$$

B.2. If $[g_1,g_2]$ denotes $g_1 g_2 g_1^{-1} g_2^{-1}$,

$$[(v_1,t_1),(v_2,t_2)] = (0, 2A_0(v_1,v_2)).$$

I.2. Discrete Subgroups of the Heisenberg Group.

Before getting into our main topic, let us pause to define a related concept, that of a rational form of the Heisenberg group. Let $N(B)$ be given and let $e_1, \ldots, e_{2n}$ be a basis of V relative to which the matrix representing B has rational entries. Then we may let $V_{\mathbb{Q}} = \sum_i r_i e_i$, $r_i \in \mathbb{Q}$, $i = 1, \ldots, 2n$, and define a group structure on $V_{\mathbb{Q}} \times \mathbb{Q}$ by the usual formula

$$(v_1, t_1)(v_2, t_2) = (v_1+v_2, t_1+t_2+B(v_1, v_2))$$

for $(v_i, t_i) \in V_{\mathbb{Q}} \times \mathbb{Q}$, $i = 1,2$. The resulting group will be called a <u>rational form</u> of $N(B)$ and denoted by $N_{\mathbb{Q}}(B)$. Clearly, $N_{\mathbb{Q}}(B) \subset N(B)$ and $N(B)$ has many rational forms which, it can be shown, are all isomorphic.

If $A_0 = \begin{pmatrix} 0 & \frac{1}{2}I_n \\ -\frac{1}{2}I_n & 0 \end{pmatrix}$ relative to the basis $e_1, \ldots, e_{2n}$. Then the basis $e_1, \ldots, e_{2n-1}, ce_{2n}$, where c is irrational will not be in any rational form of $N(A_0)$, but $ce_1, \ldots, ce_{2n}$ does determine a rational form of $N(A_0)$.

We will list without proof certain facts from [1] that we will need in these notes. Here N denotes any connected, simply connected, nilpotent Lie group.

1. Let $\Gamma \subset N$ be a discrete subgroup of N such that $\Gamma \backslash N$ is compact. Then there exists a unique form $N_{\mathbb{Q}} \subset N$ such that $\Gamma \subset N_{\mathbb{Q}}$.
2. Let $N_{\mathbb{Q}}$ be a rational form of N. Let Γ be any finitely generated subgroup of $N_{\mathbb{Q}}$. Then Γ is a discrete subgroup of N. Further, $N_{\mathbb{Q}}$ contains a finitely generated subgroup Γ such that $\Gamma \backslash N$ is compact.
3. Let Γ be a discrete subgroup of N such that $\Gamma \backslash N$ is compact. Then

$[\Gamma, \Gamma]\backslash[N, N]$ is compact, where $[\ ,\ ]$ denotes the commutator subgroup of the group in the bracket. In addition, the image of Γ in $[N, N]\backslash N$ is discrete and co-compact, i.e., has a compact homogeneous space.

4. Let Γ_1 and Γ_2 be discrete subgroups of N such that $\Gamma_1\backslash N$ and $\Gamma_2\backslash N$ are compact. Let $\psi : \Gamma_1 \to \Gamma_2$ be an isomorphism. Then ψ is uniquely extendable to an automorphism of N.

For the rest of this section we will use N to denote the $2n+1$-dimensional Heisenberg group with center $z = [N, N]$. The problem we will solve in the rest of this section is a description of all the discrete subgroups $\Gamma \subset N$ such that $\Gamma\backslash N$ is compact.

If Γ is a discrete subgroup of N such that $\Gamma\backslash N$ is compact, then $[\Gamma, \Gamma]\backslash z$ is compact. Since z is isomorphic to $\mathbb{R}$, $[\Gamma, \Gamma]$ is isomorphic to $\mathbb{Z}$. Now let $F(\Gamma) = \Gamma \cap z/[\Gamma, \Gamma]$. Note $F(\Gamma)$ is a finite cyclic group.

Definition. Let $\Gamma \subset N$ be discrete and let $\Gamma\backslash N$ be compact. We will call Γ a lifted subgroup of N provided $F(\Gamma) = e$, where e is the identity element.

The following is a description of lifted subgroups that we believe justifies their name.

Let $D \subset V$ be a discrete subgroup such that V/D is compact. We will call D a lattice subgroup of V. Let $V_{\mathbb{Q}}$ be the rational form of V containing D. Assume A is a rational nondegenerate alternating form on $V_{\mathbb{Q}}$. We may form $N_{\mathbb{Q}}(A) = V_{\mathbb{Q}} \times \mathbb{Q} \subset N(A)$. If $N(A)$ is given, then the existence of $N_{\mathbb{Q}}(A)$ is equivalent to the existence of a rational surjective $N_{\mathbb{Q}} \to V_{\mathbb{Q}}$. Let $d_1, \ldots, d_{2n}$ be generators of D and let $n_i \in N_{\mathbb{Q}}(A)$ be such that

$$p(n_i) = d_i, \qquad i = 1, \ldots, 2n,$$

where p is the homomorphism $p : N(A) \to N(A)/z = V$. Let Γ be the subgroup of $N(A)$ generated by $n_1, \ldots, n_{2n}$.

Lemma I.2.1. Γ is a lifted subgroup of $N(A)$ and $\Gamma \cap z = 2A(D \times D)$.

Proof. Since Γ is a finitely generated subgroup of $N_{\mathbb{Q}}(A)$ it is a discrete subgroup of $N(A)$. Further, by property B.2 we have if $N(A) = V \times \mathbb{R}$ and $n_i = (d_i, t_i) \in V \times \mathbb{R}$, then $[n_i, \eta_j] = [0, 2A(d_i, d_j)]$. Since A is nondegenerate and $d_1, \ldots, d_{2n}$ is a basis of V as a vector space, we have that $\Gamma \cap z$ is nonempty. Hence, $\Gamma \backslash N(A)$ is a circle bundle over the torus $D \backslash V$. It follows easily from this that $\Gamma \backslash N(A)$ is compact.

It remains to verify that $\Gamma \cap z = [\Gamma, \Gamma]$. This follows easily from the fact that any word in Γ can be written uniquely in the form

$$(d_1, t_1)^{\alpha_1} \cdots (d_n, t_n)^{\alpha_n} \delta$$

where δ is a product of commutators of the n_i. The proof of the last assertion follows from the following well-known fact [8] about alternating forms that we will state below without proof.

Lemma I.2.2. Let A be a nondegenerate alternating form on $V_{\mathbb{Q}}$. Let D be a lattice in $V_{\mathbb{Q}}$. Then we may choose a basis $d_1, \ldots, d_{2n}$ of D such that $A(d_{2j-1}, d_{2j}) = d_j \xi_0$, $j = 1, \ldots, n$, $\xi_0 \in \mathbb{Q}$, $d_j \in \mathbb{Z}^+$ and $d_j | d_{j+1}$. Further, A applied to any other pair of basis vectors is zero. Conversely, given a basis $d_1, \ldots, d_{2n}$ of $V_{\mathbb{Q}}$ and $d_j \xi_0$ as above, there exists a unique alternating form A of $V_{\mathbb{Q}}$ such that $A(d_{2j-1}, d_{2j}) = d_j \xi_0$ and A applied to any other pair of basis vectors is zero.

Lemma I.2.3. Let Γ be a discrete subgroup of $N(A)$ such that $\Gamma \backslash N(A)$ is compact. Let $D = p(\Gamma) \subset V$ and let $d_1, \ldots, d_{2n}$ be any set of generators of D. If $\gamma_i \in \Gamma$ are such that $p(\gamma_i) = d_i$, $i = 1, \ldots, 2n$, then the γ_i, $i = 1, \ldots, 2n$, generate a lifted subgroup Γ_L of Γ such that $p(\Gamma_L) = p(\Gamma)$ and $[\Gamma_L, \Gamma_L] = [\Gamma, \Gamma]$.

Proof. Since Γ is a discrete subgroup of $N(A)$ there exists a unique rational form $N_{\mathbb{Q}}(A) \supset \Gamma$. Hence, by Lemma 2.1 the γ_i, $i = 1, \ldots, 2n$, generate a discrete co-compact subgroup Γ_L of $N(A)$ and $\Gamma_L \subset \Gamma$. By the proof of Lemma 2.1 it follows that $[\Gamma, \Gamma] = [\Gamma_L, \Gamma_L]$. Since $\Gamma_L/[\Gamma_L, \Gamma_L] = \Gamma/\Gamma \cap z = D$ we have $p(\Gamma_L) = p(\Gamma)$.

Corollary I.2.4. If $[\Gamma, \Gamma] = z(N(A)) \cap \Gamma$ we have $\Gamma = \Gamma_L$.

Corollary I.2.5. Let Γ be a discrete subgroup of $N(A)$ such that $\Gamma \backslash N$ is compact. Let Γ_L be as constructed in Lemma I.2.1. Then Γ is generated as a group by Γ_L and $\xi \in z(N(A))$ and $\Gamma \cap z(N(A))$ is the group generated by ξ. Further, $[\Gamma, \Gamma] = [\Gamma_L, \Gamma_L] \subset z(N(A))$ has a generator $d\xi$, $d \in \mathbb{Z}^+$ and d is an invariant of Γ. In addition, Γ is a lifted subgroup if and only if $d = 1$.

The following definition will enable us to formulate an interesting group theoretic way of describing lifted subgroups. Any group isomorphic to

$$P = \begin{pmatrix} 1 & n_1 & n_3 \\ 0 & 1 & n_2 \\ 0 & 0 & 1 \end{pmatrix}, \qquad n_i \in \mathbb{Z},\ i = 1,2,3,$$

will be called a primitive lifted group.

Theorem I.2.6. Let Γ be a lifted subgroup of the $2n+1$ Heisenberg group. Then $\Gamma \supset \Gamma_i$, $i = 1, \ldots, n$, such that

1. Γ_i, $i = 1, \ldots, n$, are each primitive lifted groups.
2. The order of $[\Gamma, \Gamma]/[\Gamma_i, \Gamma_i]$ equals r_i, $i = 1, \ldots, n$, and $r_1 = 1$, r_i divides r_{i+1}, $i = 1, \ldots, n-1$.
3. Γ is generated by the set of subgroups Γ_i, $i = 1, \ldots, n$.
4. Γ_i centralizes each of the groups in the set $\Gamma_1, \ldots, \Gamma_n$ except itself.

Proof. Let $N(A) = \mathbb{R}^{2n} \times \mathbb{R}$ with the usual multiplication and let $p : N(A) \to \mathbb{R}^{2n}$ be the homomorphism with the center z of $N(A)$ as kernel. Let $D = p(\Gamma)$ and let $d_1, \ldots, d_{2n}$ be a basis of D satisfying the conclusions of Lemma 2.2, i.e., $2A(d_{2j-1}, d_{2j}) = \xi_j$, $j = 1, \ldots, n$, with $\xi_j \mid \xi_{j+1}$. Let $\gamma_i \in \Gamma$ be such that $p(\gamma_i) = d_i$, $i = 1, \ldots, 2n$. Then

$$[\gamma_{2j-1}, \gamma_{2j}] = 2A(d_{2j-1}, d_{2j}) = r_j[\gamma_1, \gamma_2], \quad j = 1, \ldots, n,$$

with $r_1 = 1$ and $r_j | r_{j+1}$, $j = 1, \ldots, n-1$. Further, if Γ_j is the subgroup of Γ generated by γ_{2j-1}, γ_{2j}, $j = 1, \ldots, n$, then Γ_j centralizes Γ_i for $i \neq j$. The rest of the assertions are easily verified.

Theorem I.2.7. Let $r_1, \ldots, r_n$ be elements of $\mathbb{Z}^+$ such that $r_1 = 1$ and $r_j | r_{j+1}$. Then there exists a lifted subgroup Γ and primitive subgroups

Γ_i, $i = 1, \ldots, n$, such that Γ_i satisfy the conclusions of Theorem I.2.6.

Proof. Let $d_1, \ldots, d_{2n}$ be a basis of $V_{\mathbb{Q}}$. Let A be the nondegenerate alternating form on $V_{\mathbb{Q}}$ such that $2A(d_{2j-1}, d_{2j}) = r_j$, $j = 1, \ldots, n$, and A of any other basis vectors is zero. Form $N_{\mathbb{Q}}(A) = V_{\mathbb{Q}} \times \mathbb{Q}$ and let Γ be the subgroup of $N(A)$ generated by $\gamma_\alpha = (d_\alpha, 0)$, $\alpha = 1, \ldots, 2n$. It is straightforward to verify that Γ is a lifted subgroup of $N(A)$ and if Γ_j is the subgroup of Γ generated by γ_{2j-1} and γ_{2j}, then the $\Gamma_1, \ldots, \Gamma_n$ satisfy the conclusions of Theorem I.2.6.

The content of Theorem I.2.6 may be stated in an interesting equivalent way whose proof we will leave to the reader.

Theorem I.2.6'. Let Γ be a lifted subgroup of the 2n+1-dimensional Heisenberg group. Let $P^n = P_1 \times \ldots \times P_n$ be the n-fold direct product of the primitive lifted subgroup and let $z(P^n)$ be the center of P^n. Let $k : z(P^n) \to \mathbb{Z}$ be a surjection with kernel K such that $k | z(P_i)$, $i = 1, \ldots, n$, is a monomorphism with $k(z(P_i)) \supset k(z(P_{i+1}))$, $i = 1, \ldots, n$. Then K may be chosen so that Γ is isomorphic to P^n/K.

Let Γ be a lifted subgroup of the 2n+1-dimensional Heisenberg group. If $\Gamma_1, \ldots, \Gamma_n$ and $r_1, \ldots, r_n$ satisfy the conclusion of Theorem I.2.6, we will call them a primitive presentation of Γ.

Theorem I.2.7. Let Γ and Γ' be lifted subgroups of the 2n+1-dimensional Heisenberg group and let $\Gamma_1, \ldots, \Gamma_n; r_1, \ldots, r_n$ and $\Gamma'_1, \ldots, \Gamma'_n; r'_1, \ldots, r'_n$ be primitive presentations of Γ and Γ' respectively. Then Γ is isomorphic to Γ' if and only if $r_i' = r_i$, $i = 1, \ldots, n$.

Proof. We will begin by showing that the number r_n is a group invariant for Γ.

Let Γ have center $z(\Gamma)$ which we will identify with $\mathbb{Z}$ and let $z(\ )$ denote the center of the group in the bracket. Let $q(r) : \Gamma \to \Gamma/r\mathbb{Z}$, $r \in \mathbb{Z}^+$, and let $z(\Gamma/r\mathbb{Z}) = A'(r)$. Let $p : \Gamma \to \Gamma/\mathbb{Z} \approx \mathbb{Z}^{2n}$ and let $A(r)$ denote the image of $A'(r)$ in $\mathbb{Z}^{2n}$. We assert that there exists a largest $r = r_n$ such that $A(r)$ has a subgroup which is a direct summand of $\mathbb{Z}^{2n}$. Clearly $p(\Gamma_n) \subset A(r_n)$ is a direct summand of $\mathbb{Z}^{2n}$. For $r > r_n$ it is clear that no subgroup of $A(r)$ is a direct summand of $\mathbb{Z}^{2n}$.

Let G be the maximal subgroup of $A(n)$ that is a direct summand of $\mathbb{Z}^{2n}$ and let $2k$ be its rank. Then it is easily verified that $r_{n_2} = \ldots = r_{n-1} = \ldots = r_{n-k} > r_{n-(k+1)}$. A simple induction now completes the proof of this theorem.

Theorem I.2.9. Let Γ be a discrete co-compact subgroup of $2n+1$ Heisenberg group and let Γ_L be a lifted subgroup of Γ with invariants $r_1, \ldots, r_n$. Further, that $d = \text{order}\, F(\Gamma)$. Then $d, r_1, \ldots, r_n$ are a complete system of invariants for Γ.

We will need the result of the following lemma in our proof of Theorem I.2.8.

Lemma I.2.10. Let Γ_L and Γ'_L be lifted subgroups of Γ. Then Γ_L and Γ'_L are isomorphic.

Proof. Let $\Gamma_1, \ldots, \Gamma_n$, $r_1, \ldots, r_n$, be a primitive presentation of L_L and let Γ_j have generators $\gamma_{2j-1}, \gamma_{2j}$. Let $p : \Gamma \to \Gamma/z(\Gamma)$. Then there exist $\gamma'_\alpha \in \Gamma'_L$, $\alpha = 1, \ldots, 2n$, such that $p(\gamma'_\alpha) = p(\gamma_\alpha)$. Since $\gamma'_\alpha = t_\alpha \gamma_\alpha$, $t_\alpha \in z(N)$, $\alpha = 1, \ldots, 2n$,

$$[\gamma'_\alpha, \gamma'_\beta] = [\gamma_\alpha, \gamma_\beta], \qquad \text{all } \alpha, \beta.$$

This shows that if Γ'_j has generators $\gamma'_{2j-1}, \gamma'_{2j}$, $j = 1, \ldots, n$. Then $\Gamma'_1, \ldots, \Gamma'_n$, $r_1, \ldots, r_n$, is a primitive presentation of Γ' and so by Theorem I.2.8 Γ_L is isomorphic to Γ'_L.

Proof of Theorem I.2.9. Let Γ be a discrete co-compact subgroup of N and let Γ_L be a lifted subgroup of Γ and let $\Gamma_1, \ldots, \Gamma_n$, $r_1, \ldots, r_n$, be a primitive presentation of Γ_L and let Γ_j have generators $\gamma_{2j-1}, \gamma_{2j}$. Then Γ is generated by $\gamma_0, \gamma_1, \ldots, \gamma_{2n}$ and the relations are $\gamma_0^d = [\gamma_1, \gamma_2]$ and the relations of Γ_L for $\gamma_1, \ldots, \gamma_{2n}$. This proves our assertion.

We will sometimes speak of $\gamma_0, \ldots, \gamma_{2n}$ as above as standard generators for Γ.

I. 3. The Automorphism Group of the Heisenberg Group.

In presenting the material of this section we will assume that the reader is familiar with the symplectic group, and so when we have reduced our problems to properties of the symplectic group, we will claim to have solved our problem.

Let A be a nondegenerate alternating form on V^{2n} and form the presentation of the Heisenberg group N(A). Let $\mathfrak{A}$ denote the automorphism group of N. Because of the way we have presented the Heisenberg group the 1-parameter subgroups of N(A), i.e., the continuous isomorphisms of $\mathbb{R}$ into N, are all of the form

$$t \to (t\xi, tr), \qquad t \in \mathbb{R},$$

for fixed $(\xi, r) \in N(A)$. Hence, the 1-parameter subgroups are the straight lines through the origin in the $(\xi, r) \in V^{2n} \times \mathbb{R}$ coordinate system. It follows that every $\alpha \in \mathfrak{A}$ defines a $(2n+1) \times (2n+1)$ nonsingular matrix. Since the center of N(A), z, is fixed by $\alpha \in \mathfrak{A}$, if we view elements of N(A) as the column vectors $(\xi, r)^t$, where the superscript t denotes transpose, we have

$$\alpha = \begin{pmatrix} S & 0 \\ \delta & d \end{pmatrix} . \tag{1}$$

Here S is an $n \times n$ nonsingular matrix, $\delta \in \mathbb{R}^{2n}$, and $d \neq 0$, $d \in \mathbb{R}$. Hence,

$$\alpha(\xi, r) = (S\xi, dr+\delta\xi^t) \tag{1'}$$

where $\delta\xi^t$ denotes matrix multiplication. The fact that α is an automorphism of N(A) can be expressed by the formula

$$A(S\xi, S\eta) = dA(\xi, \eta), \qquad \xi, \eta \in V^{2n}.$$

In matrix notation this becomes

(2) $$S^t A S = dA.$$

Hence, $\mathfrak{A}$ consists of all matrices of the form (1) such that S satisfy (2). Further, the subgroup of $\mathfrak{A}$ that preserves orientation of the center z, $\mathfrak{A}^0$, consists of all $\alpha \in \mathfrak{A}$ with $d > 0$.

We will now describe $\mathfrak{A}^0$ in greater detail. The symplectic group $Sp(2n, \mathbb{R})$ is defined by

$$Sp(2n, \mathbb{R}) = \{S \in GL(2n, \mathbb{R}) \mid S^t A S = A\}.$$

Define a monomorphism of $Sp(2n, \mathbb{R})$ into $\mathfrak{A}^0$ by $S \to \alpha = \begin{pmatrix} S & 0 \\ 0 & 1 \end{pmatrix}$ and identity $Sp(2n, \mathbb{R}) \subset \mathfrak{A}^0$ under this mapping. Set

(3) $$\mathcal{J} = \begin{pmatrix} I_{2n} & 0 \\ \delta^t & 1 \end{pmatrix}$$

where I_{2n} is the identity matrix and $\delta \in \mathbb{R}^{2n}$. It is easily verified that $\mathcal{J}$ is the group of inner automorphisms of $N(A)$. Equations 1' and 3 imply that we may identify $\mathcal{J}$ and $(V^{2n})^*$, where $(V^{2n})^*$ is the dual space of V^{2n}. Set

$$R = \begin{pmatrix} dI_{2n} & 0 \\ \delta^t & d^2 \end{pmatrix}, \qquad d \in \mathbb{R}.$$

Then it is easily verified that R is the radical of $\mathfrak{A}^0$ and

$$\mathfrak{A}^0 = Sp(2n, \mathbb{R}) \ltimes R .$$

We will now present an application of these concepts to a problem involving lifted subgroups of $N(A)$.

Let $D \subset V_{\mathbb{Q}}$, where $V_{\mathbb{Q}}$ has dimension $2n$, be a lattice subgroup and let A be a nondegenerate alternating form on $V_{\mathbb{Q}}$ to $\mathbb{Q}$. Consider $N_{\mathbb{Q}}(A) \subset N(A)$ and let $p : N(A) \to V$ be the homomorphism with center as kernel. Let $\Delta\{D\}$ be the set of lifted subgroups of $N(A)$ such that $\Gamma \in \Delta\{D\}$ if and only if $p(\Gamma) = D$. By our previous discussion all the elements of $\Delta\{D\}$ are isomorphic. Indeed, we may choose an isomorphism $\alpha^* : \Gamma_1 \to \Gamma_2$ such that α^* induces the identity map of D onto itself. Hence, if $\Gamma_i \in \Delta\{D\}$, $i = 1, 2$, there exists an automorphism $\alpha \in \mathfrak{A}^0$ such that $\alpha(\Gamma_1) = \Gamma_2$ and α induces the identity mapping on V. Hence, $\alpha \in \mathcal{I}$.

The following theorem is now easily verified.

Theorem I. 3. 1. Let $D \subset V_{\mathbb{Q}}$ be a lattice and let A be a nondegenerate alternating form of $V_{\mathbb{Q}}$ to $\mathbb{Q}$. Let $\Delta\{D\}$ denote the set of lifted subgroups of $N(A)$ such that for $\Gamma \in \Delta\{D\}$, $p(\Gamma) = D$. Then the group of inner automorphisms of $N(A)$, $\mathcal{I}$, acts transitively on $\Delta\{D\}$. Further, the homogeneous space $\Delta\{D\}$ inherits the structure of a compact abelian group called the dual torus of V/D.

I. 4. Fundamental Unitary Representations of the Heisenberg Group.

We will begin with a discussion of certain standard unitary representations of the Heisenberg group, relate these to induced representations and discuss the von Neumann uniqueness theorem.

Let $L^2(\mathbb{R})$ denote the space of square summable functions on the real line with Lebesgue measure. For $(x, y, t) \in \mathbb{R}^3$ and $f(s) \in L^2(\mathbb{R})$ define the unitary operator $U(x, y, t)$ by the formula

$$U(x, y, t)(f(s)) = \exp\{2\pi i\lambda(t+sy)\}f(s+x)$$

where $\lambda \in \mathbb{R}$, $\lambda \neq 0$. Now

$$\begin{aligned} U(x, y, t)\{U(a, b, c)(f(s))\} &= U(x, y, t)\{\exp(2\pi i\lambda(c+sb))f(s+a)\} \\ &= \exp(2\pi i\lambda(t+sy))\exp(2\pi i\lambda(c+(s+x)b))f(s+a+x) \\ &= U(x+a, y+b, t+c+xb). \end{aligned}$$

Hence, if we set $v_1 = (x, y)$, $v_2 = (a, b)$ we have the presentation of the 3-dimensional Heisenberg group that we called the dual pairing presentation in §1 or that given by the bilinear form with matrix

$$D = \begin{pmatrix} 0 & 1 \\ 0 & 0 \end{pmatrix}$$

and group multiplication

$$(v_1, t_1)(v_2, t_2) = (v_1+v_2, t_1+t_2+v_1Dv_2).$$

Hence, $U : (x, y, t) \to U(x, y, t)$ is a unitary representation of the 3-dimensional Heisenberg group, N. Rather than discussing induced representations in general, we will discuss this concept in the context of the Heisenberg group, N.

In N consider the subgroups $A = (0,y,t)$ and $T = (x,0,0)$. Note that A is a normal subgroup of N and N is the semidirect product of T and A which we will denote by $N = T \ltimes A$, with the conventions that the open end of the symbol $\ltimes$ always points towards the normal subgroup. The action of T on A is given by

$$(x,0,0)(0,y,t)(-x,0,0) = (0,y,t+xy).$$

Now consider all functions f on N such that

1) $f((0,y,t)(x,0,0)) = \chi(0,y,t)f(x,0,0)$, where χ is the character in A given by $\chi(0,y,t) = \exp 2\pi i\lambda t$, $\lambda \neq 0$, $\lambda \in \mathbb{R}$;

2) $f(x,0,0) \in L^2(\mathbb{R})$.

Define

$$\begin{aligned} U_1(a,b,c)f(x,0,0) &= f((x,0,0)(a,b,c)) \\ &= f((0,b,c+xb)(x+a,0,0)) \\ &= \exp(2\pi i(c+xb))f(x+a). \end{aligned}$$

Hence, $U_1 : (a,b,c) \to U_1(a,b,c)$ is the unitary representation U discussed above.

Now let $B = (x,0,t)$ and $S = (0,y,0)$. Then $N = S \ltimes B$ where the action is given by

$$(0,y,0)(x,0,t)(0,-y,0) = (x,0,t-xy).$$

Now consider the representation U_2 of N induced by

$$\chi(x,0,t) = \exp(2\pi i\lambda t).$$

Then

$$U_2(a,b,c)(f(0,y,0)) = \exp\{2\pi i\lambda(c-ay-ab)\}f(y+b).$$

Hence

$$U_2(x, y, t)U_2(a, b, c) = U_2(x+a, y+b, t+c-ay+xb).$$

Notice

$$\text{(F)} \qquad \begin{pmatrix} 0 & 1 \\ -1 & 0 \end{pmatrix}\begin{pmatrix} 0 & 1 \\ 0 & 0 \end{pmatrix}\begin{pmatrix} 0 & -1 \\ 1 & 0 \end{pmatrix} = \begin{pmatrix} 0 & 0 \\ -1 & 0 \end{pmatrix}.$$

This relation will play an important role in our discussion of the Fourier transform.

Let us now quote the following standard fundamental results:

Theorem I.4.1. For $\lambda \neq 0$, the representation

$$U_\lambda(x, y, t)f(s) = \exp\{2\pi i\lambda(t+sy)\}f(s+x)$$

is irreducible.

Theorem I.4.2. Let U_1 and U_2 be irreducible unitary representations of a group G. Assume there is a unitary operator V such that

$$VU_1V^{-1} = U_2,$$

i.e., U_1 and U_2 are unitarily equivalent; then V is unique up to multiple by the scalar operator, cI, where $|c| = 1$.

Definition. If U_1 and U_2 are unitary representations of a group G and V is such that $VU_1V^{-1} = U_2$, then V is called an intertwining operator for the representations U_1 and U_2.

Theorem I.4.3. Let U be a unitary representation of N which for each $t \in z(N)$ satisfies $U(t) = \exp\{2\pi i\lambda t\}I$, where $\lambda \neq 0$. Then U is a direct sum (finite or denumerable) of representations unitarily equivalent to U_λ.

Theorem 4.3 tells us that the unitary representations U_1 and U_2 introduced earlier in this section are unitarily equivalent. Our relation (F) suggests that there is an automorphism $\alpha : N \to N$ built from $\begin{pmatrix} 0 & 1 \\ -1 & 0 \end{pmatrix}$ such that α determines the intertwining operator between U_1 and U_2.

In the next section we will see that this is indeed the case. We will close this section by verifying directly the Fourier transform's role as an intertwining operator between U_1 and U_2 where we will choose $\lambda = 1$. Thus

$$U_1(x, y, t)(f(s)) = \exp\{2\pi i(t+sy)\}f(s+x)$$

$$U_2(x, y, t)(f(s)) = \exp\{2\pi i(c-xs-xy)\}f(s+y).$$

In carrying out this computation we will assume standard facts about the Fourier transform on $\mathbb{R}$ that can be deduced also from the computations of the next section.

Let $f(\xi) \in L^2(\mathbb{R})$. Define for $s \in \mathbb{R}$

$$\mathfrak{F}^{-1}(f)(s) = \int_{-\infty}^{\infty} f(\xi)\exp(2\pi i\xi s)d\xi.$$

Then $\mathfrak{F}^{-1}(f) \in L^2(\mathbb{R})$ and

$$\mathfrak{F}^{-1}(U_1(x,0,0)f) = \mathfrak{F}^{-1}(f(\xi+x))$$

$$= \int_{-\infty}^{\infty} f(\xi+x)\exp(2\pi i\xi s)d\xi$$

$$= \int_{-\infty}^{\infty} f(\eta)\exp(2\pi is(\eta-x)d\xi$$

$$= \exp(-2\pi isx)\mathfrak{F}^{-1}f = U_2(x,0,0)\mathfrak{F}^{-1}f \ .$$

Further,

$$\mathfrak{F}^{-1}(U_1(0,y,0)f) = \mathfrak{F}^{-1}(e^{2\pi i\xi y}f(\xi))$$

$$= \int_{-\infty}^{\infty} \exp(2\pi i\xi y)f(\xi)\exp(2\pi i\xi s)d\xi$$

$$= \mathfrak{F}^{-1}(f)(2\pi(y+s))$$

$$= U_2(0,y,0)\mathfrak{F}^{-1}f.$$

This shows that $\mathfrak{F}^{-1}U_1 = U_2\mathfrak{F}^{-1}$.

I. 5. The Fourier Transform and the Weil-Brezin Map.

We begin this section by indicating how we may extend the linear transformation $\begin{pmatrix} 0 & 1 \\ -1 & 0 \end{pmatrix}$ on V^2 to N(D). We first verify that

$$J = \begin{pmatrix} 0 & 1 & 0 \\ -1 & 0 & 0 \\ 0 & 0 & 1 \end{pmatrix}$$

operating on column vectors is an automorphism of $N(A_0)$. Let $\varphi : N(D) \to N(A_0)$ be the standard isomorphism introduced in §I.1. Then it is straightforward to verify that

$$J^* = \varphi^{-1} J \varphi \ : (x, y, t) \to (y, -x, t-xy)$$

and

$$J^{*-1} = \varphi^{-1} J^{-1} \varphi \ : (x, y, t) \to (-y, x, t-xy)$$

are the automorphisms of N(D) that we desired.

The difficulty in using J^* or J^{*-1} acting on N(D) to study intertwining operators for the representations U_1 and U_2 is that these induced representations do not live in the same Hilbert space. This problem is overcome by introducing a certain compact nilmanifold, $\Gamma\backslash N(D)$, and a subspace of functions on that nilmanifold. In this section we will use N to denote N(D).

Consider $\Gamma \subset N$, a principal subgroup of N given by $\Gamma = \{(n_1, n_2, n_3) \in N \,|\, n_i \in \mathbb{Z}\}$. Then $\Gamma\backslash N$ is a compact manifold called a nilmanifold. Consider z, the center of N acting on $\Gamma\backslash N$. Then each orbit is a circle. Indeed, $z(\Gamma)\backslash z$, where $z(\Gamma)$ denotes the center of Γ, is a circle group acting on $\Gamma\backslash N$. $\Gamma\backslash N$ is easily seen to be a principal circle bundle over $N/\Gamma z$, which is the torus $\mathbb{T}^2 = V^2/D$, where D is the lattice (n_1, n_2), $n_i \in \mathbb{Z}$, $i = 1, 2$. Since the adjoint representation of N is unipotent it follows that Haar measure on N is left and right invariant. Hence, there exists a unique Haar measure μ on N such that $\Gamma\backslash N$ has total measure one. It is easily verified that μ is given by the form

$$\mu = dx \wedge dy \wedge dt$$

in the (x, y, t) coordinate system on N. Hence, we may form $L^2(\Gamma\backslash N, \mu)$ which we will henceforth denote by $L^2(\Gamma\backslash N)$.

Since the circle group $z/z(\Gamma)$ acts on $\Gamma\backslash N$, we may decompose $L^2(\Gamma\backslash N)$ into the orthogonal direct sum

$$L^2(\Gamma\backslash N) = \oplus \sum_{n \in \mathbb{Z}} H_n(\Gamma)$$

where

$$H_n(\Gamma) = \{f \in L^2(\Gamma\backslash N) \mid f(x, y, t+s) = \exp(2\pi ins) f(x, y, t)\}.$$

The orthogonal projection, p_n, of $L^2(\Gamma\backslash N)$ onto $H_n(\Gamma)$ is given by

$$p_n(f) = \int_0^1 f(x,y, t+s) e^{-2\pi ins} ds.$$

Hence, $H_0(\Gamma)$ consists of functions that are constant on the orbits of the center z which may clearly be identified with the space $L^2(\mathbb{T}^2)$.

Now for $f \in L^2(\Gamma\backslash N)$ we define

$$\begin{aligned} R(a, b, c) f(x, y, t) &= f((x, y, t)(a, b, c)) \\ &= f(x+a, y+b, t+c+xb). \end{aligned}$$

For $f \in H_n(\Gamma)$ we have $R(a, b, c) f \in H_n(\Gamma)$. This gives us a unitary representation R_n of N on $H_n(\Gamma)$ for $n \in \mathbb{Z}$.

We will now define an intertwining operator W between U_1 and R_1. That is, we define a unitary operator $W : L^2(\mathbb{R}) \to H_1(\Gamma)$ such that $WU_1 = R_1 W$. This implies that R_1 is an irreducible unitary representation of N.

For $f(s) \in L^2(\mathbb{R})$ define the functions $f_n(\xi) = f(n+\xi)$, $0 \leq \xi < 1$, $n \in \mathbb{Z}$. Now define

$$W(f) = F(x,y,t) = \Sigma_n f_n(\xi) e^{2\pi i n y} e^{2\pi i t}$$

where $n+\xi = x$ and the sum is L^2 convergent in $\mathbb{T}^3$ where $0 \leq \xi, y, t < 1$. It is clear that the norm of $F(x,y,t)$ in $L^2(\mathbb{T}^3)$ is equal to the norm of $f \in L^2(\mathbb{R})$. Now view $W(f) = F(x,y,t)$ as a function on N. We will call $W(f)$ the Weil-Brezin mapping. We will now prove that it is invariant under the left action of Γ

$$F((n_1,n_2,n_3)(x,y,t)) = \Sigma_n f_{n+n_1}(\xi) e^{2\pi i(n+y+n_2)} e^{2\pi i(t+n_1 y)}$$

$$= \Sigma_n f_{n+n_1}(\xi) e^{2\pi i(n+n_1)y} e^{2\pi i t}.$$

Letting $n' = n+n_1$ and summing over all n' proves our assertion.

If $F(x,y,t) \in H_1(\Gamma)$, then since, up to measure zero, a fundamental domain for $\Gamma\backslash N$ is $\{(x,y,t) \in N \mid \leq y, t, t < 1\}$ and μ is Haar measure, we see that $H_1(\Gamma)$ can be naturally identified with the closed subspaces of functions of $L^2(\mathbb{T}^3)$ of the form $f(\xi_1,\xi_2)e^{2\pi i \xi_3}$, $0 \leq \xi_1, \xi_2, \xi_3 < 1$. Viewing $W(f) \in H_1(\Gamma)$, we observe that the above discussion proves that

W is an isometry. Before proving that W is surjective, let us pause to prove the following two lemmas.

Lemma I. 5. 1. Let $\mathcal{S}$ be the space of Schwarz functions on $L^2(\mathbb{R})$. Then for $f \in \mathcal{S}$

$$W(f) = \Sigma_n f(n+x)e^{2\pi iny}e^{2\pi iz}$$

where convergence is pointwise.

Proof. For each fixed x the convergence of the right-hand side follows easily from the rapid decrease of the functions in $\mathcal{S}$ as x goes to $\pm\infty$. For $0 \leq x < 1$ the lemma is obvious. It is then a formal computation to verify that if we view $W(f)$ as a function on N it is left Γ invariant. This easily implies the lemma.

Lemma I. 5. 2. Let U_1 and R_1 be as above; then if W is the Weil-Brezin map and $f \in \mathcal{S}$,

$$WU_1 f = R_1 Wf.$$

Proof. $(Wf)((x, y, t)(a, b, c)) = Wf(x+a, y+b, t+c+xb)$

$$= \Sigma_n f(x+a+n)e^{2\pi in(y+b)}e^{2\pi i(t+c+xb)}.$$

Now

$$WU_1 f = W(\exp(2\pi i(c+sb)f(s+a))$$

$$= \Sigma_n e^{2\pi iny}e^{2\pi it}\exp 2\pi i(c+(x+n)b)f(x+n+a)$$

and elementary operations verify the lemma.

Theorem I.5.3. The Weil-Brezin map $W : L^2(\mathbb{R}) \to H_1(\Gamma)$ is an isomorphism of Hilbert spaces.

Proof. It remains merely to prove that W is surjective. We have identified $H_1(\Gamma)$ with the functions in $L^2(\mathbb{T}^3)$ of the form $F = \sum_{n,m\in\mathbb{Z}} a_{nm} e^{2\pi i(n\xi_1+m\xi_2)} e^{2\pi i\xi_3}$. We may rewrite this as

$$e^{2\pi i\xi_3} \sum_n e^{2\pi i n\xi_2} \sum_m a_{nm} e^{2\pi i m\xi_1}$$

and let $f_m(\xi) = \sum_m a_{nm} e^{2\pi i m\xi_1}$ and $f(x) = f_m(\xi)$. It is clear that $f(x) \in L^2(\mathbb{R})$ and $\|f\|_{L^2(R)} = \|F\|_{L^2(\mathbb{T}^3)}$ and that $Wf = F$. This proves our theorem.

Corollary I.5.4. W is an intertwining operator for U_1 and R_1.

Corollary I.5.5. R_1 is an irreducible unitary representation of N that is unitary equivalent to U_1.

Observe: $W^{-1}(F(\xi_1, \xi_2, \xi_3)) = f(\xi_1+n) = e^{-2\pi i\xi_3} \int_0^1 F(\xi_1, \xi_2, \xi_3) e^{-2\pi i n\xi_2} d\xi_2$

Theorem I.5.6. Let $f \in L^2(\mathbb{R})$ and let J^{*-1} be the automorphism of N given by $(x, y, t) = (-y, x, t-xy)$. Then $W^{-1} J^{*-1} W : L^2(\mathbb{R}) \to L^2(\mathbb{R})$ is the Fourier transform.

Proof. Clearly, $J^{*-1} : H_1(\Gamma) \to H_1(\Gamma)$ is an isomorphism. Hence, $W^{-1} J^{*-1} W$ is an isomorphism. Therefore, the proof of this result is essentially formal at this stage and amounts to showing that

$$W^{-1}J^{*-1}Wf = \lim_{N\to\infty} \int_{-N}^{N} f(t)e^{-2\pi its}dt$$

in the L^2 norm. By our previous discussion we may work in $H_1(\Gamma)$ as though it is $e^{2\pi i\xi}{}^3L^2(\mathbb{T}^2)$ or $L^2(\mathbb{T}^2)$. Let $f(t) \in L^2(\mathbb{R})$ and $f(n+\xi) = f_n(\xi) = \sum_m a_{nm} e^{2\pi im\xi}$, $0 \leq \xi < 1$. Then $Wf = \sum_n e^{2\pi in\eta} \sum_m a_{nm} e^{2\pi im\xi}$, $0 \leq \eta$, $\xi < 1$. Notice that if $F \in L^2(\mathbb{T}^2)$, then $W^{-1}F = f(\xi+n)$

$$= \int_0^1 F(\xi,\eta)e^{-2\pi in\eta}d\eta.$$

Now
$$J^{*-1}Wf = \sum_{n,m} a_{-mn} e^{2\pi i(n\eta+m\xi)}e^{-2\pi i\xi\eta}$$

and so

$$W^{-1}J^{*-1}Wf(\xi+\alpha) = \int_0^1 \sum_{n,m} a_{-mn} e^{2\pi i(n\eta+m\xi)}e^{-2\pi i\xi\eta}e^{-2\pi i\alpha\eta}d\eta$$

$$\hat{f}(\xi+\alpha) = \sum_m \int_0^1 f_{-m}(\eta)e^{2\pi i(-\xi-\alpha)(\eta-m)}d\eta.$$

Letting $t = \eta+m$ and $s = \xi+\alpha$ we have

$$\hat{f}(s) = \lim_{N\to\infty} \int_{-N}^{N} f(t)e^{-2\pi ist}dt$$

where the limit is in the L^2 norm and the formal steps can easily be justified. Of course, letting $t' = \sqrt{2\pi}t$ and $s' = \sqrt{2\pi}s$ gives the more usual formula

$$\hat{f}(s') = \lim_{N\to\infty} \frac{1}{2\pi} \int_{-N}^{N} f(t')\exp(-is't')dt'.$$

An interesting application of the Weil-Brezin mapping is to the multiplicity problem for R_m. By Theorem I. 4. 3 we know that R_m is a multiple of the irreducible representation which on the center of N is the character $\exp(2\pi imt)$. We will now show that the multiplicity of R_m is $|m|$ using the Weil-Brezin mapping and a simple version of "left action" and distinguished subspace theory. Thus, this material serves as a motivation for our later study of distinguished subspaces and left action.

Let $\Delta_0(m)$ be the principal subgroup of the 3-dimensional Heisenberg group N given by the elements $(n_1, \frac{n_2}{m}, \frac{n_3}{m})$, where m is a fixed integer and $n_1, n_2, n_3 \in \mathbb{Z}$. Clearly, $\Delta_0(m) \supset \Gamma$. Since $\Delta_0(m)$ is principal, $H_1(\Delta_0(m))$ is irreducible and easily seen to be in $H_m(\Gamma)$. Note that an L^2 basis of $H_m(\Gamma)$ is

$$e^{2\pi imt}e^{2\pi i(\alpha x+\beta y)}, \qquad \alpha, \beta \in \mathbb{Z},$$

and an L^2 basis of $H_1(\Delta_0(m))$ is

$$e^{2\pi imt}e^{2\pi i(\alpha x+\beta my)}, \qquad \alpha, \beta \in \mathbb{Z}.$$

Let

$$\Delta_a(m) = (\frac{a}{m}, 0, 0)\Delta_0(m)(-\frac{a}{m}, 0, 0), \qquad 0 \leq a < |m|.$$

Then $\Delta_a(m) = \{(n_1, \frac{n_2}{m}, \frac{n_3}{m} + \frac{an_2}{m^2}) \in N | n_1, n_2, n_3 \in \mathbb{Z}\}$ and $\Delta_a(m)$ is a principal subgroup of N. Hence, $H_1(\Delta_a(m))$ is irreducible. We will show that

$$H_m(\Gamma) = \oplus \sum_{0 \leq a < |m|} H_1(\Delta_a(m))$$

and so establish that $H_m(\Gamma)$ is a multiplicity $|m|$ space for R_m.

Let $f \in H_m(\Gamma)$. We claim that

$$L(\frac{a}{m}, 0, 0)f = f((\frac{a}{m}, 0, 0)(x, y, t)) \in H_m(\Gamma).$$

Now $(\frac{a}{m}, 0, 0)(n_1, n_2, n_3) = (n_1, n_2, n_3 + \frac{an_2}{m})(\frac{a}{m}, 0, 0)$. But f is invariant under left action by $(n_1, n_2, n_3 + \frac{an_2}{m})$ and we have our assertion. Since $H_m(\Gamma)$ is R invariant we have

$$L(\frac{a}{m}, 0, 0)\, R(-\frac{a}{m}, 0, 0)f = f((\frac{a}{m}, 0, 0)(x, y, t)(-\frac{a}{m}, 0, 0)) \in H_m(\Gamma).$$

Now for $f \in H_1(\Delta_0(m))$ we claim that $L(\frac{a}{m}, 0, 0)R(-\frac{a}{m}, 0, 0)f \in H_1(\Delta_a(m))$. To see this, let $\delta \in \Delta_0(m)$ and $(\frac{a}{m}, 0, 0)\delta(-\frac{a}{m}, 0, 0) \in \Delta_a(m)$. Then

$$\begin{aligned} &f((\frac{a}{m}, 0, 0)\delta(-\frac{a}{m}, 0, 0)(\frac{a}{m}, 0, 0)(x, y, t)(-\frac{a}{m}, 0, 0)) \\ &\qquad = f((\frac{a}{m}, 0, 0)(x, y, t)(-\frac{a}{m}, 0, 0)). \end{aligned}$$

Hence $L(\frac{a}{m}, 0, 0)R(-\frac{a}{m}, 0, 0)H_1(\Delta_0(m)) = H_1(\Delta_a(m)) \subset H_m(\Gamma)$.

To show that $\sum_{0 \le a < |m|} H_1(\Delta_a(m))$ is an orthogonal direct sum, we consider the action of $\mathbb{Z}/m\mathbb{Z}$ on the spaces $H_1(\Delta_a(m))$ defined by $L(0, \frac{a}{m}, 0)H_1(\Delta_a(m))$. Now

$$\begin{aligned} L(0, \frac{b}{m}, 0)H_1(\Delta_a(m)) &= L(0, \frac{b}{m}, 0)L(\frac{a}{m}, 0, 0)H_1(\Delta_0(m)) \\ &= L(0, 0, -\frac{ab}{m^2})\, L(\frac{a}{m}, 0, 0)L(0, \frac{b}{m}, 0)H_1(\Delta_0(m)) \\ &= L(0, 0, -\frac{ab}{m^2})H_1(\Delta_a(m)) = \exp\text{-}2\pi i\, \frac{ab}{m} H_1(\Delta_a(m)). \end{aligned}$$

Hence each $H_1(\Delta_a(m))$ space is a character space with character

$\exp 2\pi i \frac{ab}{m}$ for the $\mathbb{Z}/m\mathbb{Z}$ group action. Therefore, they are orthogonal and the sum is direct. This shows that $\oplus \sum_{0 \leq a < |m|} H_1(\Delta_a(m)) \subset H_1(\Gamma)$.

We need only see that $H_1(\Delta_a(m))$ has

$$e^{2\pi imt} e^{2\pi i(\alpha x+(\beta m+a)y)}, \qquad \alpha, \beta \in \mathbb{Z}.$$

This is most easily done by using the observation that for $f \in H_1(\Delta_0(m))$, $L(\frac{a}{m}, 0, 0)fR(-\frac{a}{m}, 0, 0) \in H_1(\Delta_a(m))$. But $L(\frac{a}{m}, 0, 0)fR(-\frac{a}{m}, 0, 0) = f(x, y, t+\frac{a}{m}y) = \exp 2\pi iayf(x, y, t)$. Applying this to the basis for $H_1(\Delta_0(m))$ proves our assertion and the fact that

$$H_1(\Gamma) = \oplus \sum_{0 \leq a < |m|} H_1(\Delta_a(m))$$

and we have proved our multiplicity formula.

Everything in Sections I.4 and I.5 can be generalized to the 2m+1-dimensional Heisenberg group. We will close this section with a brief indication of how this is done and what mild modifications are necessary in this process.

Let $L^2(\mathbb{R}^m)$ denote the space of square summable functions in $\mathbb{R}^m$ with Lebesgue measure. For $(\underline{x}, \underline{y}, t) \in \mathbb{R}^{2m+1}$, where $\underline{x}$, $\underline{y}$ denote m-vectors and $f(\underline{s}) \in L^2(\mathbb{R}^m)$ define the unitary operator $U(\underline{x}, \underline{y}, t)$ by the formula

$$U(\underline{x}, \underline{y}, t)(f(\underline{s})) = \exp\{2\pi i\lambda(t+\underline{s} \cdot \underline{y})\}f(\underline{s}+\underline{x})$$

where $\underline{s} \cdot \underline{y}$ is the dot product $s_1y_1 + \ldots + s_ny_n$, where $\underline{s} = (s_1, \ldots, s_n)$, and $\underline{y} = (y_1, \ldots, y_n)$. Multiplication of unitary operators gives rise to a representation of the 2m+1 Heisenberg group N in the dual presentation given by

$$D = \begin{pmatrix} 0 & I_m \\ 0 & 0 \end{pmatrix}$$

where I_m is the $m \times m$ identity matrix. In the coordinate system $(\underline{x}, \underline{y}, t)$ for N multiplication is given by

$$(\underline{x}, \underline{y}, t)(\underline{x}', \underline{y}', t') = (\underline{x}+\underline{x}', \underline{y}+\underline{y}', t+t'+\underline{x}' \cdot \underline{y}').$$

In N consider the subgroups

$$A = (\underline{0}, \underline{y}, t) \qquad B = (\underline{x}, \underline{0}, t)$$

$$T = (\underline{x}, \underline{0}, 0) \qquad S = (\underline{0}, \underline{y}, t) .$$

Then $N = T \ltimes A$ and $N = S \ltimes B$. Let χ be the character in A defined by $\chi(\underline{0}, \underline{y}, t) = \exp 2\pi i\lambda t$, $\lambda \neq 0$, $\lambda \in \mathbb{R}$, and let χ' be the character on B defined by

$$\chi'(\underline{x}, \underline{0}, t) = \exp i\pi i\lambda t.$$

Then inducing χ from A to N gives the unitary representation of N given by U. Inducing χ' from B to N gives the unitary representation that satisfies

$$U_2(\underline{x}, \underline{y}, t) U_2(\underline{x}', \underline{y}', t) = U_2(\underline{x}+\underline{x}', \underline{y}+\underline{y}', t+t' - \underline{y} \cdot \underline{x}').$$

Theorems I 4.1, I 4.2 and I 4.3 hold word for word for the 2m+1-dimensional Heisenberg group and the representations U_1 and U_2 above. Let $\underline{s}, \underline{\xi} \in \mathbb{R}^m$ and $d\xi$ denote Haar measure on $\mathbb{R}^m$. Let

$$\mathcal{F}(f)(\underline{s}) = \int_{-\infty}^{\infty} f(\underline{\xi}) e^{-2\pi i \underline{\xi} \cdot \underline{s}} d\xi.$$

We have $\mathcal{F} U_2 = U_1 \mathcal{F}$.

For N_3, principal subgroups and lifted subgroups coincide. However, for the general Heisenberg group it is the principal subgroups that must be used in the Weil-Brezin map. Let Γ be a lifted subgroup of N and let us assume that coordinates in N have been chosen so that $\Gamma \cap z(N) = \mathbb{Z} = [\Gamma, \Gamma]$. Then f is an element of $H_n(\Gamma)$ if and only if

$$f(x, y, t+c) = \exp 2\pi i n c f(x, y, t).$$

We have

$$L^2(\Gamma \backslash N) = \oplus \sum_{n \in Z} H_n(\Gamma).$$

Let $\mathcal{R}_n$ be the representation of N on $H_n(\Gamma)$ given by

$$R(\underline{a}, \underline{b}, c) f(\underline{x}, \underline{y}, t) = f((\underline{x}, \underline{y}, t)(\underline{a}, \underline{b}, c)), \qquad f \in H_n(\Gamma).$$

We will use the left action and the Weil-Brezin mapping in solving the multiplicity problem for R_n in the next section.

We will close this section by outlining the Weil-Brezin map for principal subgroups P of N. Let P be the principal subgroup of N given by $(\underline{n}_1, \underline{n}_2, n_3)$, where $\underline{n}_i$ is an integer lattice point in $\mathbb{R}^m$. We will now define an isomorphism $W : L^2(\mathbb{R}^m) \to H_1(P)$ that is an inter-twining operator for U_1 and R_1.

Let D denote the integer lattice in $\mathbb{R}^m$. For $f(\underline{s}) \in L^2(\mathbb{R}^m)$ let $f_{\underline{n}}(\underline{\xi}) = f(\underline{n}+\underline{\xi})$ where $\underline{n} \in D$ and $\xi = (\xi_1, \ldots, \xi_m)$ with $0 \leq \xi_i < 1$, $i = 1, \ldots, m$. Define

$$W(f)(\underline{x}, \underline{y}, t) = \sum_{\underline{n} \in D} f_{\underline{n}}(\underline{\xi}) \exp 2\pi i(t + \underline{n} \cdot \underline{y}).$$

The results in this section for the Weil-Brezin map for $m = 1$ then go over word for word to the general case.

I.6. Distinguished Subspaces and Left Action.

In this section let N be the 2m+1-dimensional Heisenberg group in the presentation

$$(\underline{x}, \underline{y}, t)(\underline{x}', \underline{y}', t') = (\underline{x}+\underline{x}', \underline{y}+\underline{y}', t+t'+\underline{x} \cdot \underline{y}')$$

where $\underline{x}, \underline{x}', \underline{y}, \underline{y}' \in \mathbb{R}^m$ and $t, t' \in \mathbb{R}$. We will also in this section identify $\mathbb{R}$ with the center of N, z(N), and sometimes write $a \in \mathbb{R}$ to mean $(\underline{0}, \underline{0}, a) \in N$. Let Γ be a discrete subgroup of N such that $\Gamma\backslash N$ is compact and let $\beta(\Gamma) \in \mathbb{R}^+$ generate $\Gamma \cap z(N)$. Let $L^2(\Gamma\backslash N) = \oplus \sum_{n \in \mathbb{Z}} H_n(\Gamma)$ where $f \in H_n(\Gamma)$ if and only if $f \in L^2(\Gamma\backslash N)$ and

$$f(\underline{x}, \underline{y}, t+c) = \exp(2\pi i \frac{nc}{\beta(\Gamma)}) f(\underline{x}, \underline{y}, t).$$

Because in this section the group Γ will vary, we will sometimes use $R_n(\Gamma)$ to denote the regular representation of N on $L^2(\Gamma\backslash N)$ restricted to $H_n(\Gamma)$. Then $R_n(\Gamma)$ is a finite multiple of the irreducible representation $U(n/\beta(\Gamma))$ of N whose value on z(N) is a multiple of the character $\exp(2\pi i \frac{nt}{\beta(\Gamma)})$, $t \in \mathbb{R}$.

We will adopt the following conventions. $\mathcal{P}$ will denote the set of principal subgroups of N and $\mathcal{L}$ will denote the set of lifted subgroups of N. If Γ is a discrete co-compact subgroup of N we will call L_Γ a full lifted subgroup of Γ provided

1. $L_\Gamma \subset \Gamma$.
2. The image of L_Γ in N/z(N) equals the image of Γ in N/z(N).

Clearly, a full lifted subgroup of Γ is a maximal lifted subgroup of Γ.

One of the implications of Weil-Brezin map is that $R|H_1(P)$ is irreducible, where $P \in \mathcal{P}$.

Definition. Let $\mathcal{J}$ be an irreducible subspace of $H_n(\Gamma)$. We will call $\mathcal{J}$ a distinguished subspace of $H_n(\Gamma)$ if there exists $P \in \mathcal{P}$ such that $\mathcal{J} = H_1(P)$.

Lemma I.6.1. Let Γ be a discrete co-compact subgroup of N and let L_Γ be a full lifted subgroup of Γ. Then, if d is the order of Γ/L_Γ we have $H_{nd}(L_\Gamma) = H_n(\Gamma)$, $n \in \mathbb{Z}$, $n \neq 0$.

Proof. By our definition of L_Γ and $\beta(\ )$, we have that $\beta(L_\Gamma) = d\beta(\Gamma)$. Hence, functions in $H_{nd}(L_\Gamma)$ and $H_n(\Gamma)$ transform as the same central character. Further, since $L_\Gamma \subset \Gamma$,

$$H_n(\Gamma) \subset H_{nd}(L_\Gamma).$$

By Lemma I.2.2, Γ is generated by L_Γ and $\frac{1}{d}\beta(L_\Gamma)$. But if $f \in H_{nd}(L_\Gamma)$, it is also invariant under $\frac{1}{d}\beta(L_\Gamma)$. Hence $H_{nd}(L_\Gamma) = H_n(\Gamma)$.

Lemma I.6.2. Let $L_1, L_2 \in \mathcal{L}$. Then $H_r(L_1) \subset H_s(L_2)$ implies that $r\beta(L_2) = s\beta(L_1)$.

Proof. $H_r(L_1)$ is characterized as the L_1 periodic functions such that

$$f(\underline{x}, \underline{y}, t+c) = \exp(2\pi i \frac{cr}{\beta(L_1)}) f(\underline{x}, \underline{y}, t).$$

Because $f \in H_s(L_2)$, we also have that

$$f(\underline{x}, \underline{y}, t+c) = \exp(2\pi i \frac{cs}{\beta(L_2)}) f(\underline{x}, \underline{y}, t).$$

Hence

$$\frac{r}{\beta(L_1)} = \frac{s}{\beta(L_2)}$$

and we have proved our assertion.

The following remark will play an important role in the proof of the next theorem. Let $f \in H_n(\Gamma)$ and consider the usual principal circle bundle

$$\begin{array}{c} \Gamma \backslash N \\ \downarrow \pi \\ z\Gamma \backslash N = D \backslash V \end{array}$$

where z denotes the center of N and $z\Gamma$ is the subgroup of N generated by z and Γ. Then, because

$$f(x, y, t+c) = \exp(\frac{2\pi inc}{k}) f(x, y, t), \qquad k \in \mathbb{R},$$

we have that f vanishing at one point implies f vanishes on the whole fiber through the point. Hence, the set of zeros of f determines a well-defined set S^* in $D\backslash V$, or S^* in V that is D invariant. Further, let $(\underline{a}, \underline{b}, \underline{c}) \in N$ and assume that

$$f((\underline{a}, b, c)(\underline{x}, \underline{y}, \underline{t})) = f(\underline{x}, \underline{y}, t).$$

Then S^* is invariant under translation by $(\underline{a}, \underline{b})$.

Theorem I.6.3. Let $P \in \mathcal{P}$ and $L \in \mathcal{L}$. If we assume that $H_1(P) \subset H_n(L)$, then $L \subset P$.

Proof. Begin by choosing coordinates $(\underline{x}, \underline{y}, t)$ in N so that P corresponds to the points in N with integer coordinates. We will see in a later section that there exists a C^{∞} function in $H_1(P)$, say, $\varphi(\underline{x}, \underline{y}, t)$, which has its zeros on the set S described below. Let A_i, $i = 1, \ldots, m$, be the affine subspace defined by

$$A_i = \{(\underline{x}, \underline{y}, t) \mid x_i = y_i = \tfrac{1}{2}\}.$$

Let $A = \bigcup_i A_i$ and let S be the union of A and all its translates by P acting on N. Let S^* be the image of S in $V = N/z$ with coordinates $(\underline{x}, \underline{y})$. An elementary argument shows that S^* is only invariant by integer translations. This and Lemma I.6.2 then combine to prove our assertion.

Theorem I.6.4. Let $P \in \mathcal{P}$, $L \in \mathcal{L}$ and assume $H_1(P) \subset H_n(L)$. Let L have standard generators $\gamma_1, \ldots, \gamma_{2m}$ with $[\gamma_{2j-1}, \gamma_{2j}] = d_j$. Let $\Lambda(n) \in \mathcal{L}$ have standard generators $\eta_1, \ldots, \eta_{2m}$ where $\eta_{2j-1} = (d_j n)^{-1} \gamma_{2j-1}$ and $\eta_{2j} = (d_j n)^{-1} \gamma_{2j}$, $j = 1, \ldots, m$. Then $P \subset \Lambda(n)$.

Proof. Let a coordinate system be chosen in N so that

$$\gamma_{2j-1} = ((0, \ldots, 0, a_j, 0, \ldots, 0), \underline{0}, 0), \qquad j = 1, \ldots, m,$$

$$\gamma_{2j} = (\underline{0}, (0, \ldots, b_j, 0, \ldots, 0), 0).$$

Let $p \in P$ be the vector $\sum_{\alpha=1}^{2n} c_\alpha \gamma_\alpha \in N$. Then

$$[\gamma_\beta, p] = d_\beta c_\beta \geq \frac{1}{n}$$

because P has $\frac{1}{n}$ as the generator of its center. Hence $P \subset \Lambda(n)$.

Let $\Gamma(n)$ be the subgroup of N generated by L and $1/n$. Then $H_1(\Gamma(n)) = H_n(L)$. But $\Gamma(n)$ has the important algebraic property that $\Gamma(n)$ is a normal subgroup of $\Lambda(n)$, a property that L does not possess. We will use $\xi : \Lambda(n) \to \Gamma(n)\backslash\Lambda(n)$ to denote the standard homomorphism. Notice that if $H_1(P) \subset H_n(L)$, then

$$\Gamma(n) \subset P \subset \Lambda(n) \quad \text{and} \quad \beta(\Gamma(n)) = \beta(P).$$

From this it follows that $\xi(P)$ is an abelian subgroup of $\Gamma(n)\backslash\Lambda(n)$.

We will now pause to describe in a little more detail the groups $\Gamma(n)\backslash\Lambda(n)$. We will use $C(n)$ to denote the center of $\Gamma(n)\backslash\Lambda(n)$.

Let $L \in \mathcal{L}$ have standard generators $\gamma_1, \ldots, \gamma_{2n}$ with $[\gamma_{2j-1}, \gamma_{2j}] = d_j$, $d_1 = 1$, and as usual let $z(L)$ denote the center of L. Notice that $\{\gamma_{2j-1}\}$, $j = 1, \ldots, m$, and $\{\gamma_{2j}\}$, $j = 1, \ldots, m$, each generate an abelian subgroup of L which we will denote by $A(L)$ and $B(L)$, respectively. Let $\eta_1, \ldots, \eta_{2n}$ be generators of $\Lambda(n)$ as in Theorem I.6.4; i.e.,

$$\eta_{2j-1} = (d_j n)^{-1} \gamma_{2j-1}, \; \eta_{2j} = (d_j n)^{-1} \gamma_{2j}, \qquad j = 1, \ldots, m.$$

Then $A(n) = A(\Lambda(n))/A(L)$ and $B(n) = B(\Lambda(n))/B(L)$ are each isomorphic to $\sum_{j=1}^{n} \mathbb{Z}/nd_j\mathbb{Z}$.

Now $C(n) \approx \mathbb{Z}/nd_n\mathbb{Z}$ and

$$1 \to C(n) \to \Gamma(n)\backslash\Lambda(n) \to A(n) \oplus B(n) \to 1.$$

We may, and will, view $A(n)$ and $B(n)$ as subgroups of $\Gamma(n)\backslash\Lambda(n)$. Let $M(n) = C(n) \oplus A(n)$. Then

$$\Gamma(n)\backslash\Lambda(n) = B(n) \ltimes M(n).$$

<u>Theorem</u> I. 6. 5. Let all notation be as above, and let P' satisfy

$$\Gamma(n) \subset P' \subset \Lambda(n) \quad , \quad \beta(\Gamma(n)) = \beta(P).$$

Then P is principal if and only if $Z(n)$ and $\xi(P)$ generate a maximal abelian subgroup of $\Gamma(n)\backslash\Lambda(n)$.

Theorem 6. 5 has as a corollary that $H_n(\Gamma) \supset H_1(P)$, $n \neq 0$, and we have enough information to establish it directly at this time. However, we will postpone a proof of Theorem 6. 5 until after our discussion of left action. This has the advantage of proving Theorem I. 6. 5, and, considerably more, all at the same time. Although the statement of results that are to follow are slightly more general than in [2], the proofs are the same. However, to make these notes self-contained, we have included them (at least in outline form).

Let G be a separable, unimodular, locally compact group, and let Λ be a closed subgroup of G. Let Γ be a closed <u>normal</u> subgroup of Λ and assume that Γ is unimodular and that G/Γ is compact. The homogeneous space G/Γ admits a unique G-invariant probability measure that we shall not name, as it is the only measure we will consider on $\Gamma\backslash G$. On the Hilbert space $L^2(\Gamma\backslash G)$ we shall define two representations, one of G, the other of Γ/Λ:

$$(R(g)\varphi)(\Gamma h) = \varphi(\Gamma h g), \qquad \text{all } g \in G,$$

$$(L(\lambda)\varphi)(\Gamma h) = \varphi(\lambda^{-1}\Gamma h) = \varphi(\Gamma\lambda^{-1}h), \qquad \text{all } \lambda \in \Lambda,$$

for every $\varphi \in L^2(\Gamma\backslash G)$. As we have written above, L is a representation of Λ, but because Γ is the kernel of L, we can (and usually shall) view L as a representation of $\Gamma\backslash\Lambda$.

Let $(\Gamma\backslash G)^\wedge$ denote the set of unitary equivalence classes of irreducible unitary representations of G that occur as subrepresentations of the representation R. It is a classical result that $L^2(\Gamma\backslash G)$ decomposes into an orthogonal direct sum $\oplus \Sigma H_p(\Gamma)$, the sum being taken over every $p \in (\Gamma\backslash G)^\wedge$, and each $H_p(\Gamma)$ is an R-invariant subspace such that $R|H_p(\Gamma)$ is a finite multiple of p. Another classical result asserts that the orthogonal projection of $L^2(\Gamma\backslash G)$ onto $H_p(\Gamma)$ lies in the von Neumann algebra generated by the operators $\{R(g), g \in G\}$. Because $L(\lambda)R(g) = R(g)L(\lambda)$ all $g \in G$, it follows that the operators $L(\lambda)$ commute with the orthogonal projection of $L^2(\Gamma\backslash G)$ onto $H_p(\Gamma)$. In other words, each of the subspaces $H_p(\Gamma)$ is L-invariant as well as R-invariant.

Since $R|H_p(\Gamma)$ is the equivalence class k_p for some integer $k > 0$, if we fix an irreducible R-invariant subspace $\mathfrak{H}_0 \subset H_p(\Gamma)$, there is an isometric isomorphism

$$K : H_p(\Gamma) \to \mathfrak{H}_0 \otimes \mathbb{C}^k$$

such that

$$(R(g) \otimes 1)K\varphi = KR(g)\varphi, \qquad g \in G \text{ and } \varphi \in H_p(\Gamma).$$

<u>Lemma</u> I.6.6. There is a unitary representation L_p of $\Gamma\backslash\Lambda$ on $\mathbb{C}^k$ such that for all $\lambda \in \Gamma\backslash\Lambda$ and all $\varphi \in H_p(\Gamma)$

$$KL(\lambda)\varphi = (1 \otimes L_p(\lambda))K\varphi .$$

Furthermore, L_P does not depend, up to unitary equivalence, on the choice of $\mathfrak{H}_0$ and K.

Proof. Because $R|\mathfrak{H}_0$ is irreducible, the only bounded operators on $\mathfrak{H}_0 \otimes \mathbb{C}^k$ that commute with all of the operators $(R(g)|\mathfrak{H}_0) \otimes 1$ have the form $1 \otimes T$ for some $T : \mathbb{C}^k \to \mathbb{C}^k$. Hence

$$KL(\lambda)K^{-1} = 1 \otimes L_p(\lambda)$$

for some $L_p(\lambda) : \mathbb{C}^k \to \mathbb{C}^k$. The rest of the assertions are standard.

We will now specialize all this to our case of $\Gamma(n) \subset \Lambda(n) \subset N$, where N is the Heisenberg group.

Let η be the character on $C(n) \simeq \mathbb{Z}/nd_n\mathbb{Z}$, the center of $\Gamma(n)\backslash\Lambda(n)$ given by $\eta(a) = \exp(2\pi i\, a/d_n n)$, $a \in \mathbb{Z}/nd_n\mathbb{Z}$. Clearly, the restriction of $R_n(\Gamma)$ to $C(n)$ is a multiple of η. Hence the restriction of L_n to $C(n)$ is also some multiple of η.

Theorem I.6.7. There is, up to unitary equivalence, precisely one irreducible representation $T(\Gamma, n)$ of $\Gamma(n)\backslash\Lambda(n)$ whose restriction to $C(n)$ is some multiple of η. The dimension of $T(\Gamma, n) = d_1 \ldots d_m |n|^m$.

Proof. Let $\Gamma(n)\backslash\Lambda(n) = B(n) \ltimes M(n)$ as discussed earlier in this section. Let W be the set of characters of $M(n)$ that extend η. If $\zeta \in W$ and $b \in B(n)$, then for $a \in M(n)$

$$a \to \zeta(b^{-1}ab)$$

is a character. Hence $B(n)$ acts as a group of transformations on W.

One checks that the isotropy group for $\zeta \in W$ is trivial. Since the number of characters of $M(n)$ extending η is the same as the number of elements of $B(n)$, $B(n)$ acts simply transitively on W.

Let T be the representation induced from $M(n)$ and the character $\zeta \in W$ to $\Gamma(n)\backslash\Lambda(n)$. It is easily checked that $T|M(n)$ is of the form

$$\begin{pmatrix} \zeta_1 & & 0 \\ & \ddots & \\ 0 & & \zeta_N \end{pmatrix}$$

where N = order of $B(n) = d_1 \ldots d_m |n|^m$ and $\zeta_1, \ldots, \zeta_N$ are the elements of W. Further, $B(n)$ acts as the shift operators. Clearly, T is irreducible and the dimension of T is the order of $B(n) = d_1 \ldots d_m |n|^m$.

Now suppose that S is a second irreducible representation of $\Gamma(n)\backslash\Lambda(n)$ restricting to a multiple of η on $C(n)$. We must show that S and T are equivalent. Clearly, it suffices to prove that S is a subrepresentation of T. Since T is induced by ζ, Frobenius reciprocity says that S is a subrepresentation of T if ζ is a subrepresentation of $S|M(n)$. Now there is certainly at least one character $\omega \in W$ that occurs in $S|M(n)$. But, for $b \in B(n)$ the character $a \to \omega(b^{-1}ab)$ of $M(n)$ must occur in $S|M(n)$, because S is equivalent to the representation of $\Gamma(n)\backslash\Lambda(n)$ given by $\gamma \to S(b^{-1}\gamma b)$. Hence ω occurring in $S|M(n)$ implies that ζ occurs in $S|M(n)$ as required.

Theorem I.6.8. $L(n)$ is equivalent to $T(\Gamma, n)$.

Proof. Assume the coordinates have been chosen so that $\beta(P) = 1$, and let P^* be the group generated by P and $z(N)$. Let χ denote the character of P^* given by

$$(\underline{a}, \underline{b}, c) \rightarrow \exp 2\pi i n c.$$

One requires the well-known result that $R_n(P)$ is equivalent to the representation $X(n)$ of N obtained by inducing χ from P^* to N.

We next use the theorem on inducing by stages, i.e., if $A \supset B \supset C$ and we induce C to B and B to A, then this is the same as inducing C to A. Let $M^*(n)$ be the group generated by $\xi^{-1}(M(n))$ and $z(N)$. We denote by Q the representative of $M^*(n)$ induced by χ. Then

$$Q = \oplus \sum_{i=1}^{N} \omega_i .$$

The sum over all the $N = d_1 \ldots d_m |n|^m$ distinct characters ω_i of $M^*(n)$ whose restriction to P^* is χ. Since the representation of N induced by Q is equivalent to $X(n)$ on the one hand and to the sum of induced representations $\oplus\Sigma\, I(\omega_i)$, where $I(\omega_i)$ is the representation obtained by inducing ω_i to N, the crux of the proof is in showing that $I(\omega_i)$ is irreducible. It will then follow from the Stone-von Neumann theorem that $I(\omega_i)$ is equivalent to $U(n)$. Since $M^*(n)$ is normal in N and χ is fixed only by $M^*(n)$, it follows from Mackey that $I(\omega_i)$ is irreducible.

Corollary I.6.8. Given L there exists $P \in \mathcal{P}$ such that $H_1(P) \subset H_n(L)$.

Proof. Let η be as in Theorem 6.7. Let $\zeta \in W$ be a character extending η. Let $A(\zeta) \subset M(n)$ be the kernel of ζ. Then $M(n) = A(\zeta) \oplus C(n)$. Then it is easily seen that $P = \xi^{-1}(A(\zeta))$ is principal, and let V be the one-dimensional subspace of $\mathbb{C}^k$ on which $A(\zeta)$ acts trivially. Let $\mathfrak{J} \subset H_n(L)$ correspond to V. Then $H_1(P) = \mathfrak{J}$.

We note that in Corollary I.6.8 we made an arbitrary choice where we wrote $M(n) = A(\zeta) \oplus C(n)$, so that P really should be denoted by $P(\zeta)$. Clearly, as ζ runs through W we get $H_1(P(\zeta)) \subset H_n(L)$.

Corollary I.6.9. $H_n(L) = \oplus \sum_{\zeta \in W} H_1(P(\zeta))$.

Proof. $B(n)$ acts on W simply transitively. But $B(n)$ acts on $H_n(L)$ by left action. Clearly,

$$R(b)L(b)H_1(P(\zeta)) = H_1(P(b^{-1}\zeta b)).$$

This proves our assertion.

We will now look at the corollaries in a more general setting.

Let $\mathfrak{J}$ be an $R_n(L)$ invariant subspace of $H_n(L)$. We define the left stabilizer $\Lambda(\mathfrak{J})$ of $\mathfrak{J}$ by

$$\Lambda(\mathfrak{J}) = \{\lambda \in \Gamma(n)\backslash\Lambda(n) \mid L(\lambda)\mathfrak{J} = \mathfrak{J}\}.$$

Theorem I.6.10. $\Lambda(\mathfrak{J})$ is an abelian subgroup of $\Gamma(n)\backslash\Gamma(n)$ and there is a character ζ of $\Lambda(\mathfrak{J})$ such that $L(\lambda)\varphi = \zeta(\lambda)\varphi$ for all $\varphi \in \mathfrak{J}$ and $\lambda \in \Lambda(\mathfrak{J})$.

Proof. Let $K : H_n(L) \to \mathfrak{J}_0 \otimes \mathbb{C}^N$, $N = d_1 \dots d_m |n|^m$, be the usual isometric isomorphism. Let V be the one-dimensional subspace of $\mathbb{C}^N$ such that $K(\mathfrak{J}) = \mathfrak{J}_0 \otimes V$. Then

$$\Lambda(\mathfrak{J}) = \{\lambda \in \Gamma(n)\backslash\Lambda(n) \mid L_n(\lambda)V = V\}.$$

Hence there is a character ζ of $\Lambda(\mathfrak{J})$ such that whenever $\lambda \in \Lambda(\mathfrak{J})$ and $v \in V$ we have $L_n(\lambda)v = \zeta(\lambda)v$. Clearly, $C(n) \subset \Lambda(\mathfrak{J})$ and restricting ζ to $C(n)$ is a faithful character.

Now $[\Lambda(\mathfrak{J}), \Lambda(\mathfrak{J})] \subset C(n)$ and also in the kernel of ζ. Hence $\Lambda(\mathfrak{J})$ is abelian.

In [2] the following two results are proven even if they are not stated.

Theorem I.6.11. Let L be principal and let Δ be an abelian subgroup of $\Gamma(n)\backslash\Lambda(n)$, $\Delta \supset C(n)$. For each character ζ of Δ such that $\Delta \mid C(n) = \eta$ set

$$H_n(L,\zeta) = \{\varphi \in H_n(L) \mid L(\delta) = \zeta(\delta)\varphi \text{ for all } \delta \in \Delta\}.$$

Then

1. $H_n(L,\zeta)$ is R-invariant.
2. The multiplicity of $U(n)$ in $R \mid H_n(L,\zeta)$ is $d_1 \ldots d_m |n|^{m+1}/\text{order}\,\Delta$.
3. $H_n(L) = \oplus \Sigma H_n(L,\zeta)$, the sum over all characters ζ of Δ such that $\zeta \mid C(n) = \eta$.

Corollary I.6.12. $H_n(L,\zeta)$ is irreducible if and only if Δ is maximal abelian.

Lemma I.6.13. Let Δ be an abelian subgroup of $P(n)\setminus\Lambda(n)$. Then the order of Δ, denoted by $|\Delta : 1|$ divides $|n|^{m+1}$ and Δ is maximal abelian if and only if $|\Delta : 1| = |n|^{m+1}$.

Definition. Let $\mathcal{J}$ be an R invariant irreducible subspace of $H_n(P)$, $P \in \mathcal{P}$. We define the index of $\mathcal{J}$, denoted by $\mathrm{ind}(\mathcal{J})$ to be $|n|^{m+1}/|\Lambda(\mathcal{J}) : 1|$.

Lemma I.6.14. Let A be an abelian subgroup of $\Gamma(n)\setminus\Lambda(n)$ such that $A \cap C(n) = e$. Then the intersection of the conjugates of A is the identity.

Proof. For $a \in A$ there exists $d \in \Gamma(n)\setminus\Lambda(n)$ such that

$$d^{-1}ad = a+c\ ,\quad c \in C(n),\qquad c \neq C.$$

The lemma follows easily from this fact.

Theorem I.6.15. Let $P \in \mathcal{P}$ and $L \in \mathcal{L}$ and assume that

$$H_r(P) \subset H_s(L). \quad \text{Then } L \subset P.$$

Proof. By Theorem I.6.11 and Lemma I.6.14

$$H_r(P) = \oplus\Sigma H_1(P_\alpha)$$

where $\cap P_\alpha = \Gamma(r)$. Then $\oplus\Sigma H_1(P_\alpha) \subset H_s(L)$, and so by Theorem I.6.3 $P_\alpha \supset L$ all α, and so $\cap P_\alpha \supset L$. This implies that $P \supset L$.

Before ending this section, I should relate the version of distinguished subspaces presented here with the version used in [3] and [11]. In the previous references the structure of distinguished subspaces is given in terms of the automorphism groups of N. We may relate this by noting that if $P_1, P_2 \in \mathcal{P}$ and $\psi : P_1 \to P_2$ is an isomorphism there exists a unique automorphism ψ^* of N such that $\psi^* | P_1 = \psi$. Now let $P_0 \in \mathcal{P}$ and P_1 be such that $H_1(P_1) \subset H_n(P_0)$. Then there exists an automorphism of N taking P_0 to P_1. This observation gives the technique for relating our work to that in [3] and [11].

II. Jacobi Theta Functions and the Finite Fourier Transform.

II.1. Nil-Theta Functions and Jacobi Theta Functions.

In this section we will begin our study of nil-theta functions and their relation to the classical Jacobi-theta functions. There are two approaches to establishing the results we desire. One can assume complex variable results and use these to establish properties of nil-theta functions. Or one can directly prove properties of nil-theta functions and use these to obtain classical results. We will mainly follow the first approach in this chapter. In Chapter III we will mostly follow the second approach.

In this Chapter we will restrict our attention to the 3-dimensional Heisenberg group N_3 and use the presentation

$$(x_1, y_1, t_1)(x_2, y_2, t_2) = (x_1+x_2, y_1+y_2, t_1+t_2+\tfrac{1}{2}(y_1x_2-x_1y_2))$$

or using $\mathbb{C} \times \mathbb{R}$ and $(z_i, t_i) \in \mathbb{C} \times \mathbb{R}$, $i = 1,2$,

$$(z_1, t_1)(z_2, t_2) = (z_1+z_2, t_1+t_2+\tfrac{1}{2}\,\mathrm{Im}\, z_1\bar{z}_2).$$

Recall that in N_3 every lifted subgroup is principal.

Given the elliptic Riemann surface $E(\tau) = \mathbb{C}/L(\tau)$, where

$$L(\tau) = \{k_1+k_2\tau \in \mathbb{C} \mid k_i \in \mathbb{Z},\ i = 1,2,\ \tau = \alpha+i\beta,\ \beta > 0\}.$$

Jacobi showed that one of the most powerful tools for the study of the field of meromorphic functions on $E(\tau)$, or elliptic functions, is theta functions with characteristics. In this section we will show the deep relation between this classical theory and functions on a nilmanifold. An explanation for the relationship will be offered in Chapter III.

In N_3 let $\Gamma(\tau)$ be the principal subgroup generated by $(1,0,0)$ and $(\alpha,\beta,0)$. Then

$$\Gamma(\tau) = \{(k_1+k_2\alpha,\ k_2\beta,\ k_3\beta-\tfrac{1}{2}k_1k_2\beta) \in N_3 \mid k_i \in \mathbb{Z},\ i = 1,2,3\}.$$

Further, $z(\Gamma(\tau)) = (0,0,\beta k_3)$. Hence the characters on $z(N_3)/z(\Gamma(\tau))$ are $\exp(2\pi i m t\beta^{-1})$, $m \in \mathbb{Z}$. Again

$$\begin{array}{c}\Gamma(\tau)\backslash N_3 \\ \downarrow \\ L(\tau)\backslash \mathbb{C}\end{array}$$

is a principal circle bundle with the orbits of $z(N_3)/z(\Gamma(\tau))$ as fibers.

Changing notation slightly, we have

$$L^2(\Gamma(\tau)\backslash N_3) = \oplus \sum_{m\in V} H_m(\tau).$$

We let $C_m^\infty(\tau) = C^\infty(\Gamma(\tau)\backslash N_3) \cap H_m(\tau)$ and $C_m^\infty(\tau)$ is dense in $H_m(\tau)$. (Here our functions are complex valued!) Clearly, if $F_m \in C_m^\infty(\tau)$, then $F_mF_q \in C_{m+q}^\infty(\tau)$.

Of course, we will view elements of $C^\infty(\tau)$ as $\Gamma(\tau)$ periodic functions on N_3 whenever this suits our convenience.

Let us pause to give a slightly different formulation of the Cauchy-Riemann equations. Consider $\mathbb{C}$ as a real vector space $\mathbb{R}^2$ the correspondence given by $x+iy \to (x,y)$. Then multiplication by i corresponds to the linear transformation $J((x,y)) = (-y,x)$. Now view $\mathbb{R}^2$ as a Lie group and $\frac{\partial}{\partial x}, \frac{\partial}{\partial y}$ as a basis for the left invariant vector fields on $\mathbb{R}^2$. Then the Lie algebra of $\mathbb{R}^2$, complexified, $L_{\mathbb{C}}(\mathbb{R}^2) = \{c_1 \frac{\partial}{\partial x} + c_2 \frac{\partial}{\partial y} \mid c_i \in \mathbb{C},\ i = 1, 2\}$. For $f \in C^\infty(\mathbb{R}^2)$, $f = f_1 + i f_2$, f_i real valued

$$\frac{\partial}{\partial x} f = \frac{\partial f_1}{\partial x} + i \frac{\partial f_2}{\partial x}$$

$$\frac{\partial}{\partial y} f = \frac{\partial f_1}{\partial y} + i \frac{\partial f_2}{\partial y}.$$

Now $V_{-i} = \frac{\partial}{\partial x} + i \frac{\partial}{\partial y} = \frac{\partial}{\partial \bar{z}}$ is an eigenvector for J with eigenvalue $-i$. Hence, $f \in C^\infty(\mathbb{R}^2)$ is analytic if and only if $V_{-i}(f) = 0$. Thus we may view $\mathbb{C}$ as the pair $(\mathbb{R}^2, J)$ and $\frac{\partial}{\partial \bar{z}} = V_{-i}$.

Now the complex numbers $\mathbb{C} = (\mathbb{R}^2, J)$ gives rise to an automorphism J of N_3 given by

$$J(x, y, t) = (-y, x, t).$$

Now let $L(N_3)$ be the Lie algebra of N_3. Viewing $L(N_3)$ as left invariant vector fields on N_3, we have a basis of $L(N_3)$ is given by the vector fields X, Y, T

$$X(x, y, t) = \frac{\partial}{\partial x} + \tfrac{1}{2} y \frac{\partial}{\partial t} \Big|_{(x, y, t)}$$

$$Y(x, y, t) = \frac{\partial}{\partial y} - \tfrac{1}{2} x \frac{\partial}{\partial t} \Big|_{(x, y, t)}$$

$$T(x, y, t) = \frac{\partial}{\partial t} \Big|_{(x, y, t)}$$

where $(x, y, t) \in N_3$.

Since J is an automorphism of N_3 it induces an automorphism J_* of $L(N_3)$ given by

$$J_*(X) = -Y , \quad J_*(Y) = X , \quad J_*(T) = T.$$

Now consider $L_{\mathbb{C}}(N_3)$ the complexification of $L(N_3)$. Then for $f \in C^\infty(\Gamma(\tau)\backslash N_3)$ and $W \in L_{\mathbb{C}}(N_3)$ we have $Wf \in C^\infty(N_3)$ is invariant under $\Gamma(\tau)$. This is because for $\gamma \in \Gamma(\tau)$, $(Wf)(\gamma n) = W(\gamma n)f(\gamma n) = W(n)f(n)$. Now in $L_{\mathbb{C}}(N_3)$ let V_{-i} be an eigenvector with eigenvalue $-i$. Then $X+iY = kV_{-i}$, $k \in \mathbb{R}$. Hence we may choose

$$V_{-1} = \left(\frac{\partial}{\partial x}+i\frac{\partial}{\partial y}\right) - \tfrac{1}{2}(x+iy)\frac{\partial}{\partial t} = \frac{\partial}{\partial \bar{z}} - \tfrac{1}{2}z\frac{\partial}{\partial t}.$$

Definition. We define the space of nil-theta functions on N_3, Θ_N, as the subspace of $C^\infty(N_3)$ such that $F \in \Theta_N$ if and only if $V_{-i}F = 0$.

Let Γ be a principal subgroup of N_3. The V_{-i} is a well-defined operator on $C^\infty(\Gamma\backslash N_3)$. Let $\Theta_N(\Gamma) = C^\infty(\Gamma\backslash N_3) \cap \Theta_N$.

Lemma II.1.1. Let p_n be the orthogonal projection of $L^2(\Gamma\backslash N_3)$ onto $H_n(\Gamma)$, $n \neq 0$. Let $F \in \Theta_N(\Gamma)$. Then $p_n(F) \in \Theta_N(\Gamma)$.

Proof. $V_{-i}(p_n F)(x,y,t) = \int_0^{\beta(\Gamma)} \exp\left(\frac{-2\pi i m s}{\beta(\Gamma)}\right) V_{-i}(F(x,y,t+s))ds$ from which the lemma follows immediately.

Define $\Theta_n(\Gamma) = \Theta_N(\Gamma) \cap C_n^\infty(\Gamma)$.

The left action discussed in I.6 proves useful here in obtaining a picture of $\Theta_n(\Gamma)$.

Lemma II.1.2. Let $\lambda \in \Gamma(n)\backslash\Lambda(n)$. Then $L_n(\lambda)\Theta_n(\Gamma) = \Theta_n(\Gamma)$.

Proof. $0 = V_{-i}F$, $F \in \Theta_n(\Gamma) \subset H_n(\Gamma)$. Then

$$0 = L_n(\lambda)(V_{-i}F) = V_{-i}(L_n(\lambda)F).$$

Notice if $\mathcal{J}$ is an irreducible subspace of $H_n(\Gamma)$ and $\Theta_n(\Gamma)$ is not the zero element, then $\mathcal{J} \cap \Theta_n(\Gamma)$ is not empty. Further, if $\Theta_n(\Gamma)$ is finite-dimensional, then $\dim \Theta_n(\Gamma) = |n| C$ where C is some positive integer or ∞.

Because we will use the exponential function so often, we will sometimes adopt the notation $e(a)$ to denote e^a.

The following result will prove essential in relating nil-theta functions to the classical Jacobi theta functions.

Theorem II. 1. 3. Let $F \in C^\infty(N_3)$ be such that $V_{-i}F = 0$ and

$$F(x, y, t) = e^{(2\pi iat)} F(x, y, 0), \qquad a \in \mathbb{R}.$$

Then

$$M(a)F = e^{(-2\pi iat)} e^{(-\pi iaxy)} e^{(\pi a y^2)} F(x, y, t)$$

is an entire function of $z = x+iy$.

Conversely, let $\xi(x, y)$ be an entire function of $z = x+iy$. Then

$$M^{-1}(a)\xi(x, y) = e^{(2\pi iat)} e^{(\pi iaxy)} e^{(-\pi a y^2)} \xi(x, y)$$

is a solution of V_{-i}.

Proof. We have to show that if $V_{-i}F = 0$ and $F(x, y, t) = e^{(2\pi iat)} F(x, y, 0)$, $a \in \mathbb{R}$, then $M(a)F$ satisfies the Cauchy-

Riemann equations

$$\frac{\partial}{\partial \bar{z}} M(a)F = 0.$$

Let $A(x,y) = e(-\pi iaxy)e(\pi ay^2)$ and $\varphi(x,y) = F(x,y,0)$. Then we need to show that $A\varphi$ satisfies the Cauchy-Riemann equations. Since $V_{-i}F = 0$, we have

$$\frac{\partial}{\partial \bar{z}} \varphi(x,y) = -a\pi z\varphi(x,y).$$

Hence

$$\frac{\partial}{\partial \bar{z}}(A\varphi) = (\frac{\partial A}{\partial \bar{z}} - a\pi zA)\varphi.$$

But, by an elementary computation one verifies that

$$\frac{\partial A}{\partial \bar{z}} = a\pi zA,$$

and this verifies the first half of our theorem.

Let $\xi(x,y)$ be an entire function. We must verify that $V_{-i}(M(a)^{-1}\xi(x,y)) = 0$. First note that

$$V_{-i}(M(a)^{-1}\xi) = M(a)^{-1}V_{-i}(\xi) + \xi V_{-i}(M(a)^{-1})$$

and that $\frac{\partial \xi}{\partial \bar{z}} = \frac{\partial \xi}{\partial t} = 0$. An elementary computation proves that $V_{-i}(M(a)^{-1}) = 0$. These facts combine to prove that $V_{-i}(M(a)^{-1}\xi) = 0$.

Corollary II.1.4. Let P be the principal subgroup of N_3. Then $\Theta_m(P)$ is trivial for $m < 0$ and $\Theta_0(P)$ consists of the constant functions.

Proof. Let $F(x, y, t) \in C^{\infty}(N_3)$ and P-periodic. Clearly, $F(x, y, t)$ is a bounded function. Now $e(\pi a y^2)$ is a bounded function for $a \leq 0$. Hence, $M(a)F$ is a bounded entire function and hence a constant. Since there are no nontrivial constant functions in $H_n(P)$, $n \neq 0$, we have our result.

Corollary II.1.5. Let $f, g \in \Theta_n(\Gamma(\tau))$, $n > 0$, then f/g is a meromorphic function on $\mathbb{C}/L(\tau)$.

Theorem II.1.6. Let $F \in \Theta_n(\Gamma(\tau))$, $n > 0$, $\tau = \alpha + i\beta$, $\beta > 0$. Let $z = x + iy$ and $\xi(z) = M(\frac{n}{\beta})F(x, y, t)$, $n > 0$. Then

1. $\xi(z+1) = \xi(z)$,
2. $\xi(z+\tau) = e(-\pi i n(2z+\tau))\xi(z)$.

In other words, $\Theta_n(\Gamma(\tau))M(\frac{n}{\beta})$ are the classical Jacobi theta functions of period τ and characteristic $[{0 \atop 0}]$. These are usually denoted by $\Theta_n[{0 \atop 0}](z, \tau)$.

Conversely, let $\xi(z)$ be entire and satisfy 1 and 2. Then

$$(M(\tfrac{n}{\beta})^{-1}\xi)(x, y, t) \in \Theta_n(\Gamma(\tau)).$$

Proof. Since F is $\Gamma(\tau)$ periodic we have

$$F(x, y, 0) = F((1, 0, 0)(x, y, 0)) = F(x+1, y, -\tfrac{1}{2}y)$$

$$= e(\frac{-2\pi i m y}{2\beta})F(x+1, y, 0)$$

and

$$F(x, y, 0) = F((\alpha, \beta, 0)(x, y, 0)) = F(x+\alpha, y+\beta, \tfrac{1}{2}(\beta x - \alpha y))$$

$$= e(\frac{2\pi i m}{2\beta}(\beta x - \alpha y))F(x+\alpha, y+\beta, 0).$$

Recall that $M(\frac{m}{\beta})(x, y, t) = e(\frac{-2\pi imt}{\beta})e(\frac{-mixy}{\beta})e(\frac{\pi my^2}{\beta})$. This enables us to compute as follows:

$$(M(\frac{m}{\beta})F)(z+1) = e(\frac{-\pi im(x+1)y}{\beta})e(\frac{\pi my^2}{\beta})F(x+1, y, 0)$$

$$= e(\frac{-\pi imy}{\beta})M(\frac{m}{\beta})e(\frac{\pi imy}{\beta})F(x, y, 0)$$

$$= (M(\frac{m}{\beta})F)(z).$$

Next

$$(M(\frac{m}{\beta})F)(z+\tau) = e(\frac{-\pi im}{\beta}(x+\alpha)(y+\beta))e(\frac{\pi m}{\beta}(y+\beta)^2)$$

$$= e(\frac{-\pi im}{\beta}(\beta x - \alpha y))F(x, y, 0)$$

which after some algebraic manipulations yields

$$(M(\frac{m}{\beta})F)(z+\tau) = e(-\pi im(2z+\tau))(M(\frac{m}{\beta})F)(z).$$

The converse is proven by completely analogous reasoning.

A re-examination of the proof of Theorem II. 1. 6 shows that it is a purely formal proof and so we obtain the following

Corollary II. 1. 7. $F \in C_m^\infty(\tau)$ if and only if $M(\frac{m}{\beta})F$ satisfies equations 1 and 2 of Theorem II. 3. 4.

Corollary II. 1. 8. Let $\mathcal{J} \subset H_n(\Gamma(\tau))$, $n > 0$, be R-invariant and irreducible. Then $\dim J \cap \Theta_n(\Gamma(\tau)) = 1$.

Proof. By classical results one knows that $\dim \Theta_n(\Gamma(\tau)) = n$, $n > 0$. The corollary now follows easily from Lemma II. 1. 2.

Let $\oplus \sum_{n=1}^{\infty} \Theta_n(\Gamma(\tau))$ denote the algebraic direct sum, i. e., all but a finite number of terms are zero. Then we will make $\sum_{n=1}^{\infty} \Theta_n(\Gamma(\tau))$ into an algebra by considering it as a subspace of $C^\infty(\Gamma(\tau)\backslash N_3)$. Let $\oplus \sum_{n=1}^{\infty} \Theta_n[{0 \atop 0}](z,\tau)$ denote the algebraic direct sum of vector spaces of Jacobi theta functions. Make $\sum_{n=1}^{\infty} \Theta_n[{0 \atop 0}](z,\tau)$ into an algebra by considering it as a subspace of the algebra of entire functions on the complex plane $\mathbb{C}$ and proving it is a subalgebra.

Theorem II. 1. 9. The isomorphisms $M(\frac{n}{\beta})\Theta_n(\Gamma(\tau)) = \Theta_n[{0 \atop 0}](z,\tau)$ extends to an algebra isomorphism $M : \oplus \Sigma\Theta_n(\Gamma(\tau)) \to \oplus \Sigma\Theta_n[{0 \atop 0}](z,\tau)$.

Corollary II. 1. 10. The algebra

$$\Sigma\Theta_n(\Gamma(\tau))$$

has no divisors of zero.

Proof of Theorem II. 1. 9. This amounts to verifying that

$$M(\frac{m}{\beta})M(\frac{m'}{\beta}) = M(\frac{m+m'}{\beta})$$

which is a formal computation.

In the beginning of this section we selected a particular lifted subgroup $\Gamma(\tau)$ over the lattice $L(\tau) \subset \mathbb{C}$. Using $\Gamma(\tau)$, we related $\Theta_n(\Gamma(\tau))$ to the classical Jacobi theta functions $\Theta_n[{0 \atop 0}](z,\tau)$. It is reasonable to ask the following question: Given $\Gamma'(\tau)$ a lifted subgroup over $L(\tau)$, do the spaces $\Theta_n(\Gamma'(\tau))$ correspond to classical objects. We will now show that this is indeed the case.

By Theorem I. 3. 1 we know that the group of inner automorphisms of N_3 acts transitively on all the lifted subgroups over $L(\tau)$. Thus given $\Gamma(\tau)$, $\tau = \alpha + i\beta$, $\beta > 0$, there exists $g \in N_3$ such that

$$\Gamma'(\tau) = g^{-1}\Gamma(\tau)g.$$

For reasons that will become clear as a result of later computation, we let $g = (a+b\alpha, b\beta, 0)$, $a, b \in \mathbb{R}$. If $\gamma = (k_1+k_2\alpha, k_2\beta, c) \in \Gamma(\tau)$

$$g^{-1}\gamma g = (k_1+k_2\alpha,\ k_2\beta,\ c+(b\beta k_1 - a\beta k_2))$$

and we will denote $g^{-1}\Gamma(\tau)g$ by $\Gamma(\tau;a,b)$.

<u>Theorem</u> II.1.11. There exist $\Gamma(\tau;a,b)$ periodic nil-theta functions.

<u>Proof.</u> We define $Ad(\tau;a,b)F(x,y,t) = F(g^{-1}(x,y,t)g)$. If F is $\Gamma(\tau)$-periodic, we have

$$G(x,y,t) = Ad(\tau;a,b)F(x,y,t)$$

is $\Gamma(\tau;a,b)$ invariant. Let $F \in C_m^\infty(\tau)$. Then

$$Ad(\tau;a,b)F \in C_m^\infty(\Gamma(\tau;a,b))$$

and

$$G(x,y,t) = e\left(\frac{2\pi i m}{\beta}(b\beta x - (a+b\alpha)y\right)F(x,y,t).$$

We will now examine the functional equations that $M(\frac{m}{\beta})G$ satisfies. Since F satisfies equations 1 and 2 above, we have that

3. $(M(\frac{m}{\beta})G)(z+1) = e(2\pi imb)(M(\frac{m}{\beta})G)(z).$
4. $(M(\frac{m}{\beta})G)(z+\tau) = e(-2\pi ima)e(-\pi im(2z+\tau))(M(\frac{m}{\beta})G)(z).$

By classical results, we know that there exist entire functions that satisfy equations 3 and 4.

We may now reason as above to prove

Theorem II.1.12. $M(\frac{n}{\beta})\Theta_n(\tau;a,b)$, $n > 0$, is the space of Jacobi theta functions of periodic τ, order n and characteristic $[{a \atop b}]$. Let $\mathcal{J} \subset H_n(\Gamma(\tau;a,b))$, $n > 0$, be an R-invariant, irreducible subspace. Then $\mathcal{J} \cap \Theta_m(\Gamma(\tau;a,b))$ has dimension one.

We will now introduce two bases of $\Theta_n(i)$ that will play an important role in a later application. Consider $H_n(i)$ and Γ' be generated by $(1,0,0)$ and $(0,\frac{1}{n},0)$ and Γ'' generated by $(\frac{1}{n},0,0)$ and $(0,1,0)$. Since the image of Γ' and Γ'' in $\Gamma(n)\backslash\Lambda(n)$ and the center of $\Gamma(n)\backslash\Lambda(n)$ generate a maximal abelian subgroup of $\Gamma(n)\backslash\Lambda(n)$, it follows that $H_1(\Gamma') \subset H_n(i)$ and $H_1(\Gamma'') \subset H_n(i)$.

Now let $\Gamma'_a = (\frac{a}{n},0,0)\Gamma'(-\frac{a}{n},0,0)$ and $\Gamma''_a = (0,\frac{a}{n},0)\Gamma''(0,-\frac{a}{n},0)$. Then Γ'_a is generated by $(1,0,0)$ and $(0,\frac{1}{n},-\frac{a}{n^2})$ and Γ''_a is generated $(\frac{1}{n},0,\frac{a}{n^2})$ and $(0,1,0)$. We wish to explicitly find $\Theta_n(i) \cap \Gamma'_a$ and $\Theta_n(i) \cap \Gamma''_a$.

Because of the fundamental domains of Γ' and Γ'', we are seeking theta functions in $\mathbb{C}$ corresponding to the analytic tori $\mathbb{C}/L'$ and $\mathbb{C}/L''$ where $L' = \{(k_1,k_2/n) \mid k_i \in \mathbb{Z},\ i=1,2\}$ and $L'' = \{(k_1/n,k_2) \mid k_i \in \mathbb{Z},\ i=1,2\}$. Now, using classical notation $\theta[{0 \atop 0}](\zeta,i/n)$ and $\theta[{0 \atop 0}](n\zeta,n_i)$ are first order theta functions for the analytic tori $\mathbb{C}/L'$ and $\mathbb{C}/L''$ respectively, where explicitly

$$\theta[{0 \atop 0}](\zeta,i/n) = \Sigma_m e(\pi i(\frac{i}{n}m^2+2m\zeta))$$

$$\theta[{}^0_0](n\zeta, ni) = \Sigma_m e(\pi i(nim^2+2nm\xi)).$$

Since for $\Gamma(i)$, $\beta = 1$, we have

$$M(n)^{-1} = e(2\pi int)e(\pi nxy)e(-\pi ny^2)$$

and $$M(n)^{-1}\theta[{}^0_0](\zeta, i/n) \in H_1(\Gamma') = \theta'_{n0}(x,y,t) \in H_n(i)$$

and $$M(n)^{-1}\theta[{}^0_0](n\zeta, ni) \in H_1(\Gamma'') = \theta''_{n0}(x,y,t) \in H_n(i).$$

To see explicitly that $\theta[{}^0_0](\zeta, i/n)$ and $\theta[{}^0_0](n\zeta, ni)$ lie in $\Theta_n[{}^0_0](\zeta, i)$, we have to verify that they satisfy the required functional equations. We of course know that

$$\theta[{}^0_0](\zeta+1, \tau) = \theta[{}^0_0](\zeta, \tau)$$

$$\theta[{}^0_0](\zeta+\tau, \tau) = e(\pi i(-2\zeta-\tau))\theta[{}^0_0](\zeta, \tau).$$

Hence $$\theta[{}^0_0](n(\zeta+\tfrac{1}{n}), ni) = \theta[{}^0_0](n\zeta+1, ni) = \theta[{}^0_0](n\zeta, ni)$$

and
$$\begin{aligned}\theta[{}^0_0](n(\zeta+i), ni) &= \theta[{}^0_0](n\zeta+ni, ni)\\ &= e(\pi i(-2n\zeta-ni))\theta[{}^0_0](n\zeta, ni)\\ &= \exp \pi in(-2\zeta-i)\theta[{}^0_0](n\zeta, ni).\end{aligned}$$

Thus $$\theta[{}^0_0](n\zeta, ni) \in \Theta_n[{}^0_0](\zeta, i).$$

The argument for $\theta[{}^0_0](\zeta, i/n)$ is even simpler and will be omitted.

Let $L(\frac{a}{n}, 0, 0)\theta'_{n0} = \theta'_{na}$ and $L(0, \frac{a}{n}, 0)\theta''_{n0} = \theta''_{na}$. Then $\theta'_{na} \in H_1(\Gamma'_a)$ and $\theta''_{na} \in H_1(\Gamma''_a)$ and $\theta'_{n0}, \ldots, \theta'_{nn-1}$ and $\theta''_{n0}, \ldots, \theta''_{nn-1}$ are each orthogonal basis of $\Theta_n(i)$. Since left multiplication is a unitary operator, every element in each basis has the same norm. Further, since we have explicit formulas for θ'_{n0} and θ''_{n0} it is straightforward to obtain explicit formulas for θ'_{na} and θ''_{na}.

II.2. The Algebra of the Finite Fourier Transform.

In this section we will give an application of the material that we have developed to this point. We will see how the material we have developed enables us to piece together the finite Fourier transforms for $\mathbb{Z}/n\mathbb{Z}$ for all n to form an algebra.

Again let N_3 be the 3-dimensional Heisenberg group with the presentation

$$(x_1, y_1, t_1)(x_2, y_2, t_2) = (x_1+x_2, y_1+y_2, t_1+t_2+\tfrac{1}{2}(y_1x_2-x_1y_2)).$$

Let $\Gamma(i) = \Gamma_0$ be the principal subgroup of N_3 generated by $(1,0,0)$ and $(0,1,0)$. Then if $\gamma \in \Gamma_0$

$$\gamma = (k_1, k_2, k_3-\tfrac{1}{2}k_1k_2).$$

Consider $\Gamma_0\backslash N_3$. Crucial to this section is that

$$J(x, y, t) \to (-y, x, t)$$

an automorphism of N_3, has the additional property that $J(\Gamma_0) = \Gamma_0$. To see this, note that

$$J(1,0,0) = (0,1,0) \quad \text{and} \quad J(0,1,0) = (-1,0,0).$$

Since $(-1,0,0)$ is the inverse of $(1,0,0)$, it is obvious that Γ_0 is also generated by $(-1,0,0)$ and $(0,1,0)$. We will obtain our result from the fact that $J(\Gamma_0\backslash N_3) = \Gamma_0\backslash N_3$ and that J determines the operator V_{-i}.

Let us begin the more formal part of this section with a brief discussion of certain finite nilpotent groups that we will call mod-n

Heisenberg groups. Perhaps the simplest definition of the mod-n Heisenberg groups are given as follows: The group Γ_0 has the faithful representation

$$\begin{pmatrix} 1 & n_1 & n_3 \\ 0 & 1 & n_2 \\ 0 & 0 & 1 \end{pmatrix}$$

$n_i \in \mathbb{Z}$, $i = 1,2,3$. We define the mod-n Heisenberg group, denoted by $\Gamma/(n)$ by reducing each of the n_i, $i = 1,2,3$, mod-n. Hence $\Gamma/(n)$ is the matrix group over the ring $\mathbb{Z}/n\mathbb{Z}$ of the form

$$\begin{pmatrix} 1 & a_1 & a_3 \\ 0 & 1 & a_2 \\ 0 & 0 & 1 \end{pmatrix} \qquad a_i \in \mathbb{Z}/n\mathbb{Z},\ i = 1,2,3.$$

Hence $\Gamma(n)$ has the presentation

$$(a_1, a_2, a_3)(b_1, b_2, b_3) = (a_1+b_1, a_2+b_2, a_3+b_3+a_1b_2)$$

where $a_i, b_i \in \mathbb{Z}/n\mathbb{Z}$. It follows easily that $\Gamma/(n)$ satisfies the exact sequence

$$1 \to \mathbb{Z}/n\mathbb{Z} \to \Gamma/(n) \xrightarrow{h} \mathbb{Z}/n\mathbb{Z} \oplus \mathbb{Z}/n\mathbb{Z} \to 1$$

where the kernel of h is the center of $\Gamma/(n)$. We will now give a description of $\Gamma/(n)$ in terms of group extensions. Let $\mathcal{Z}(\Gamma/(n))$, the center of $\Gamma/(n)$, be identified with $e(2\pi i \frac{a_3}{n})$, $0 \leq a_3 < n$, and let $(a_1, a_2) \in \mathbb{Z}/n\mathbb{Z} \oplus \mathbb{Z}/n\mathbb{Z}$. Let $\gamma_1, \gamma_2 \in \Gamma/(n)$ be such that $h(\gamma_1) = (1,0)$ and $h(\gamma_2) = (0,1)$. If the symbol $[\ ,\ ]$ denotes the commutator of the elements in the bracket, then the group extension $\Gamma/(n)$ is completely determined by $[\gamma_1^{a_1}, \gamma_2^{a_2}]$. An elementary computation then yields that $[\gamma_1^{a_1}, \gamma_2^{a_2}] = e(2\pi i \frac{a_1 a_2}{n})$. We see therefore that the group $\Gamma/(n)$ is built from the pairing of $\mathbb{Z}/n\mathbb{Z}$ and its character group to the circle group given by the duality of finite abelian groups.

Let us now see what the general facts given in Section I. 6, particularly Theorem I. 6. 7, imply in our special case. Let η be the faithful character on $\mathfrak{Z}(\Gamma/(n))$ given by

$$\eta(a_3) = \exp(2\pi i \frac{a_3}{n}), \qquad a_3 \in \mathfrak{Z}(\Gamma/(n)).$$

Let

$$B = (a_1, 0, a_3) \qquad S = (0, a_2, 0)$$

$$A = (0, a_2, a_3) \qquad T = (a_1, 0, 0) \ .$$

Then $\Gamma/(n) = T \ltimes A$ and $\Gamma/(n) = S \ltimes B$. Let ζ be the character of A extending η given by $\zeta(a_1, 0, a_3) = \eta(a_3)$ and let ζ' be the character of B extending η given by $\zeta'(0, a_2, a_3) = \eta(a_3)$. Then by Theorem I. 6. 7, $\mathrm{Ind}(\zeta, A)$ and $\mathrm{Ind}(\zeta', B)$ are unitarily equivalent. Sections I. 4 and I. 5 suggest that the finite Fourier transform should serve as an intertwining operator for $\mathrm{Ind}(\zeta, A)$ and $\mathrm{Ind}(\zeta', B)$. Further, the automorphism $J(n)$ of $\Gamma/(n)$ given by

$$J(n)(a_1, a_2, a_3) = (-a_2, a_1, a_3)$$

should be related to the finite Fourier transform. We will now develop these ideas in some detail.

Note the action of T on A is given by

$$(a_1, 0, 0)(0, a_2, a_3)(-a_1, 0, 0) = (0, a_2, a_3 - a_1 a_2)$$

and the action of S on B is given by

$$(0, a_2, 0)(a_1, 0, a_3)(0, -a_2, 0) = (a_1, 0, a_3 + a_1 a_2).$$

Now consider all functions $f \in L^2(\Gamma/(n))$ such that

$$f((0, a_2, a_3)(a_1, 0, 0)) = \zeta(0, a_2, a_3) f(a_1, 0, 0)$$

$$= \exp(2\pi i \frac{a_3}{n}) f(a_1, 0, 0).$$

The above set of functions is clearly a vector subspace which we will denote by $\mathcal{F}_T$. Note we may map

$$\mathcal{I}_T : L^2(\mathbb{Z}/n\mathbb{Z}) \to \mathcal{F}_T$$

by letting $\mathcal{I}_T(h(a_1))$, $h(a_1) \in L^2(\mathbb{Z}/n\mathbb{Z})$ be defined by

$$\mathcal{I}_T(h(a_1)) = \zeta(0, a_2, a_3) h(a_1) = f((0, a_2 a_3)(a_1, 0, 0)).$$

Note $\|\mathcal{I}_T(h)\| = n\|h\|$. Then $\mathrm{Ind}(\zeta, A)$ is given by

$$\mathrm{Ind}(\zeta, A)(b_1, b_2, b_3) f(a_1, 0, 0) = f((a_1, 0, 0)(b_1, b_2, b_3))$$

$$= f((0, b_2, b_3 - a_1 b_2)(a_1 + b_1, 0, 0))$$

$$= \exp \frac{2\pi i}{n} (b_3 - a_1 b_2) f(a_1 + b_1, 0, 0).$$

Notice this corresponds to what we called the U_2 representative of the Heisenberg group.

Now consider functions on $L^2(\Gamma/(n))$ such that

$$f((a_1, 0, a_3)(0, a_2, 0)) = \exp(2\pi i \frac{a_3}{n}) f(0, a_2, 0)$$

and denote this vector space by $\mathfrak{F}_S$. This time we map

$$\mathcal{S}_S : L^2(\mathbb{Z}/n\mathbb{Z}) \to \mathfrak{F}_S$$

by $\quad \mathcal{S}_S(h(a_2)) = \exp(2\pi i \frac{a_3}{n}) h(a_2) = f((a_1, 0, a_3)(0, a_2, 0))$.

Then

$$\operatorname{Ind}(\zeta', B)(b_1, b_2, b_3) f(0, a_2, 0) = \exp \frac{2\pi i}{n}(b_3 + a_2 b_1) f(0, a_2 + b_2, 0).$$

(This corresponds to the unitary representation of the Heisenberg group that we called U_1.)

We already know that the analogues of Theorems I. 4. 1 and I. 4. 2 and I. 4. 3 hold in our finite setting. Let us now see how the finite Fourier transform serves as an intertwining operator. For $f \in L^2(\mathbb{Z}/n)$

$$\mathfrak{F}_n(f)(s) = \frac{1}{\sqrt{n}} \sum_{\xi=0}^{n-1} f(\xi) e^{\frac{-2\pi i \xi s}{n}}.$$

To show that $\mathfrak{F}_n \operatorname{Ind}(\zeta, A) = \operatorname{Ind}(\zeta', B) \mathfrak{F}_n$ one computes as follows.

$$\begin{aligned} \mathfrak{F}_n(\operatorname{Ind}(\zeta, A)(b_1, 0, 0)f) &= \frac{1}{\sqrt{n}} \sum f(\xi + b_1) e(-2\pi i \frac{\xi s}{n}) \\ &= e(2\pi i \frac{s b_1}{n}) \mathfrak{F}_n(f) \\ &= \operatorname{Ind}(\zeta', B)(b_1, 0, 0) \mathfrak{F}_n f, \end{aligned}$$

etc.

In Section I. 5 we showed that the Fourier transform could be viewed as the automorphism J operating on $H_1(\Gamma_0\backslash N_3)$. In this section we will see that the finite Fourier transform $\mathfrak{F}_n$ can be viewed as operating on $\Theta_n(i)$, $n > 0$.

Let $\Gamma_0(n)\backslash\Lambda(n)$ be the group of left actions on $\Theta_n(i)$, $n > 0$. We know that $\Gamma_0(n)\backslash\Lambda(n) \approx \Gamma/(n)$ acts irreducibly on $\Theta_n(i)$.

Theorem II. 2. 1. $J\Theta_n(i) = \Theta_n(i)$, $n > 0$.

Proof. J determines an automorphism of $L_{\mathbb{C}}(N_3)$ and $J(V_{-i}) = cV_{-i}$, $c \neq 0$. Let $f \in \Theta_n(i)$. Then

$$0 = V_{-i}(f), \text{ and so } 0 = J(V_{-i}f) = cV_{-i}(Jf).$$

Note in Section II. 1 we proved

1. $\dim\Theta_n(i) = n$, $n > 0$.

2. $\oplus\,\Sigma_{n>0}\Theta_n(i)$ is an algebra without divisors of zero. Again let $\Gamma(n) = T \ltimes A$ and $S \ltimes B$ and let η be the character of $\mathfrak{z}(\Gamma/(n))$ given by $\eta(a_3) = \exp(2\pi i\frac{a_3}{n})$.

Now J operates on $\Theta_n(i)$ as a linear transformation which we will denote by J_n. Further, J determines an automorphism of $\Gamma(n)\backslash\Lambda(n)$ because $J(\Gamma(n)) = \Gamma(n)$ and $J(\Lambda(n)) = \Lambda(n)$. Let L_n denote the left action of $\Gamma(n)\backslash\Lambda(n)$ on $\Theta_n(i)$. Then L_n is irreducible. Let $L_{n_J} = L_n \circ J_n$. Then L_{n_J} is irreducible and J acts trivially on the center of $\Gamma(n)\backslash\Lambda(n)$, L_{n_J} is unitarily equivalent to L_n.

Because L_n is unitary equivalent to $\mathrm{Ind}(\zeta, A)$, we may choose a unitary operator U_n on $\Theta_n(i)$ such that

$$U_n^{-1} L_n U_n^{-1} = \mathrm{ind}(\zeta, A).$$

It is easy to see that

$$U_n^{-1} L_{n_J} U_n = \mathrm{Ind}(\zeta', B).$$

Hence, there exists an orthogonal coordinate system in $\Theta_n(i)$, $e_1, \ldots, e_n$, such that L_n and L_{n_J} written as matrices relative to this basis are $\mathrm{Ind}(\zeta, A)$ and $\mathrm{Ind}(\zeta', B)$. Hence, relative to this basis one intertwining operator is the matrix $\mathfrak{F}_n$. By a straightforward computation $J_n L_n J_n^{-1} = L_{n_J}$. Hence, $J_n = c\,\mathfrak{F}_n$, $c = \pm 1, \pm i$. To see that $c = 1$ requires a computation using distinguished subspaces that we will indicate below.

Consider $H_n(\Gamma_0) = \oplus \sum_{a=0}^{n-1} H_1(\Gamma_a)$,

where $$\Gamma_a = \{(\frac{n_1}{n}, n_2, \frac{n_3}{n} + \frac{an_1}{n^2}) \mid n_i \in \mathbb{Z},\ i = 1,2\}$$

and $$H_n(\Gamma_0) = \oplus \sum_{a=0}^{n-1} H_1(\Gamma'_a)$$

$$\Gamma'_a = \{(n_1, \frac{n_2}{n}, \frac{n_3}{n} + \frac{an_2}{n^2}) \mid n_i \in \mathbb{Z},\ i = 1,2\}.$$

We observe that $\theta'_{na} \in \Theta_1(\Gamma_a)$ and $\theta''_{na} \in \Theta_1(\Gamma'_a)$, θ'_{na} and θ''_{na} as in II.1. Now it is easily seen that θ''_{n0} has its n zeros in $0 \leq x, y \leq 1$ at the points $\zeta = a + \frac{1}{2}i$, where $a = \frac{1}{2n}, \frac{1}{2n} + \frac{1}{n}, \ldots, \frac{2n-1}{2n}$,

$$\zeta = \frac{2a-1}{2n} + \tfrac{1}{2}i, \qquad a = 1, \ldots, n \quad ,$$

and θ'_{n0} has its zero at

$$\zeta = \tfrac{1}{2} + \frac{2a-1}{2n} i, \qquad a = 1, \ldots, n \quad .$$

Since an entire function is determined up to a scalar multiple by its zeros, it follows that $J(\theta''_{n0}) = c_1 \theta'_{n0}$. To determine c_1, we note that $J(0,0,0) = (0,0,0)$ and that both $\theta''_{n0}(0,0,0)$ and $\theta'_{n0}(0,0,0)$ are positive at $(0,0,0)$. Thus $c_1 > 0$.

Next $\mathfrak{F}_n \theta''_{n0} = \Sigma_{a=0}^{n-1} \theta''_{aa}$. By a direct computation based on the explicit formulas of the last section, we have $\mathfrak{F}_n \theta''_{n0} = (\Sigma \theta''_{na})(0,0,0) > 0$. Since $\mathfrak{F}_n \theta''_{n0} = c\, J_n \theta''_{n0}$ and both $(\mathfrak{F}_n \theta''_{n0})(0,0,0) > 0$ and $(J_n \theta''_{n0})(0,0,0) > 0$, we have $c > 0$ and so $c = 1$. We have now proved

Theorem II.2.2. Let $\oplus \Sigma_{n>0} \Theta_n(i) = \mathfrak{A}$. Define $\mathfrak{F}_n$ on $\Theta_n(i)$ as J_n of the basis θ''_{na}, $a = 0, \ldots, n-1$, then $\mathfrak{F}(fg) = \mathfrak{F}(f)\,\mathfrak{F}(g)$, $f, g \in \mathfrak{A}$, and $\mathfrak{A}$ has no zero divisors.

Let us identify $\Theta_n(i)$ and $L^2(\mathbb{Z}/n\mathbb{Z})$ by assigning to θ''_{na} the function taking the value one at a and zero elsewhere. Notice we have identified $\mathbb{Z}/n\mathbb{Z}$ and $(\mathbb{Z}/n\mathbb{Z})^\wedge$ by $a \to \exp 2\pi i \frac{ab}{n}$, $a, b \in \mathbb{Z}/n\mathbb{Z}$, in order to obtain the matrix representation of $\mathfrak{F}_n$ by

$$\mathfrak{F}_n = (\exp 2\pi i \frac{ab}{n}).$$

In [4] Tolimieri and I establish the following facts that I will just quote here.

Let $$L = \oplus \Sigma_{n>0} L^2(\mathbb{Z}/n\mathbb{Z})$$

and let $\mathfrak{F}_n$ denote the Fourier transform on $L^2(\mathbb{Z}/n)$ after we have identified $\mathbb{Z}/n\mathbb{Z}$ and $(\mathbb{Z}/n\mathbb{Z})^\wedge$ by $a \to \exp 2\pi i \frac{ab}{n}$, $a, b \in \mathbb{Z}/n\mathbb{Z}$. Let $\mathfrak{F}$ denote the linear transformation of L such that $\mathfrak{F} | L^2(\mathbb{Z}/n\mathbb{Z}) = \mathfrak{F}_n$, all n.

<u>Theorem</u> II. 2. 3. L has, up to isomorphism, three structures as an algebra L_α, $\alpha = 1, 2, 3$, such that

1. if $f \in L^2(\mathbb{Z}/n)$ and $g \in L^2(\mathbb{Z}/m)$, $f{,}g \in L^2(\mathbb{Z}/n{+}m)$;
2. L_α, $\alpha = 1, 2, 3$, has no divisors of zero;
3. for $f, g \in L_\alpha$, $\alpha = 1, 2, 3$

$$\mathfrak{F}(fg) = \mathfrak{F}(f)\,\mathfrak{F}(g).$$

Further,

$$L_1 = \mathbb{C}[X_1, X_2^2, X_3^3]/(X_3^6+X_2^6)$$

$$L_2 = \mathbb{C}[X_1, X_2^2, X_3^3]/X_2^6+X_1^4X_2^2+X_2^6)$$

$$L_3 = \mathbb{C}[X_1, X_2^2, X_3^3]/(X_3^6+X_1^4X_2^2).$$

<u>Theorem</u> II. 2. 4. $\oplus \Sigma \Theta_n(i)$ is isomorphic to L_2.

III. Nil-Theta Functions and Abelian Varieties.

III.1. A General Construction and Some Algebraic Foundations.

Let M be a compact manifold and let $H^*(M, \mathbb{R})$ and $H^*(M, \mathbb{Z})$ be the cohomology rings of X with real and integer coefficients respectively. Let $H^+(M, \mathbb{R})$ and $H^+(M, \mathbb{Z})$ denote the subring all of whose elements have positive degree. Then $H^+(M, \mathbb{R})$ is a nilpotent associative algebra. Let $1 \in H^0(M, \mathbb{R})$ be the identity of the ring $H^*(M, \mathbb{R})$ and let

$$N(M) = \{1+a \,|\, a \in H^+(M, \mathbb{R})\}.$$

Then $N(M)$ is a subgroup of the group of units of the algebra $H^*(H, \mathbb{R})$, and it is easily seen that $N(M)$ is a connected, simply connected nilpotent Lie group. We will set

$$N^j(M) = \{1+a \,|\, a \in H^j(M, \mathbb{R}),\ j > 0\} \subset N(M).$$

In general, $N^j(M)$ is not a group. Now the Lie algebra of $N(M)$, $L(M)$, is formed from the associative algebra $H^+(H, \mathbb{R})$ by defining $[a, b] = ab-ba$, $a, b \in H^+(H, \mathbb{R})$. Let $a \in H^\alpha(M, \mathbb{R})$ and $b \in H^\beta(M, \mathbb{R})$; then $a \cap b = (-1)^{\alpha\beta}$ $b \cap a \in H^{\alpha+\beta}(M, \mathbb{R})$. If $a \in H^\alpha(M, \mathbb{R})$, α even, a commutes with all $b \in H^+(M, \mathbb{R})$. Hence, $\Sigma_{\alpha > 1} H^{2\alpha}(M, \mathbb{R}) \subset L(M)$ is the center of $L(M)$, $\mathfrak{z}(L(M))$. Further, $[L(M), L(M)] \subset \mathfrak{z}(L(M))$ and so $L(M)$ and $N(M)$ are 2-step nilpotent.

Now by the standard theory of cohomology rings, we have a natural injection $H^*(M, \mathbb{Z}) \xrightarrow{i} H^*(M, \mathbb{R})$ such that $i(H^j(M, \mathbb{Z}))$ is a lattice in the vector space $H^j(M, \mathbb{R})$, $j > 1$. Consider

$$\Gamma(M) = \{1+n \mid n \in i(H^+(M, \mathbb{Z}))\}.$$

Then $(1+n_1)(1+n_2) = 1+n_1+n_2+n_1n_2$. Since $n_1+n_2+n_1n_2 \in i(H^+(M, \mathbb{Z}))$ it follows that $\Gamma(M)$ is closed under multiplication. That $\Gamma(M)$ is a subgroup follows from the fact that

$$(1+n)^{-1} = 1-n+n^2 - \dots \pm n^N.$$

Hence, $\Gamma(M)$ is a discrete subgroup of $N(M)$ and it is easily seen that $\Gamma(M)\backslash N(M)$ is compact. Hence, we have functorially assigned to every compact manifold M a compact nilmanifold $\Gamma(M)\backslash N(M)$.

Now let M be a complex manifold. Then A. Weil in [14] shows that the complex structure on M determines an automorphism J_M of $H^*(M, \mathbb{R})$. Further, if $f : M \to N$ is a complex analytic mapping, then

$$f^* J_N = J_M f^*.$$

Hence, the complex structure determines an automorphism of $N(M)$ which we will also denote by J_M.

Now let S be a Riemann surface of genus $m > 0$. Then it is classical that $N(S)$ is the $2m+1$-dimensional Heisenberg group and that J_S on $N(S)/\mathfrak{z}(N(S))$ has the property that its square is minus the identity and J_S acts as the identity on $\mathfrak{z}(N(S))$.

Now consider a complex torus $\mathbb{C}^m/L = T$, where L is a discrete co-compact subgroup of $\mathbb{C}^m$. Then $H^*(\mathbb{C}^m/L, \mathbb{R}) = \Lambda^*(V^*)$, where $\Lambda^*(\ )$ denotes the exterior algebra and V^* is the dual to the real vector space of dimension 2m underlying $\mathbb{C}^m$. Further, J_T on $\Lambda^1(V^*)$ is the automorphism described below and J_T acts trivially on $\Lambda^0(V^*)$. Since $\Lambda^0(V^*)$ and $\Lambda^1(V^*)$ generates $\Lambda^*(V^*)$ as an algebra, knowing $J_T|\Lambda^1(V^*)$ completely describes $\Lambda^*(V^*)$. Consider $\mathbb{C}^m$ as the real vector space V. Then multiplication by i in $\mathbb{C}^m$ determines a linear transformation J of V such that $J^* = -I$. Clearly, J determines an automorphism J^* of V^*. Identifying $\Lambda^1(V^*)$ with V^* we have $J_T|\Lambda^1(V^*) = J^*(V^*)$.

Now let T be the Jacobi variety of S and let $\varphi : S \to T$ be the complex mapping of the Riemann surface S into its Jacobi variety. Then we have $\varphi^* : N(T) \to N(S)$ and $\varphi^* : \Gamma(T) \to \Gamma(S)$. <u>We will now abstract the properties of the mapping φ^* and show the fundamental role in the theory of abelian varieties and theta functions of mappings with such properties.</u>

Let us begin by getting a better understanding of J_S.

<u>Lemma</u> III.1.1. Let A be an automorphism of the Heisenberg group such that if the induced action of A on $N/\mathfrak{z}$ is denoted by A_1, then $A_1^2 = -I$ and let $V \times \mathbb{R}$ be the basic presentation of N. Then there exists an inner automorphism n of N such that

$$n^{-1}An = \begin{pmatrix} A_1 & 0 \\ 0 & 1 \end{pmatrix}.$$

<u>Proof.</u> By the structure theory of automorphisms of N given in Chapter I, we have $\mathfrak{A}^0 = Sp(2n, \mathbb{R}) \ltimes \mathcal{R}$. Hence

$$A = \begin{pmatrix} A_1 & 0 \\ 0 & 1 \end{pmatrix} \begin{pmatrix} dI & 0 \\ \delta & d^2 \end{pmatrix}.$$

Since $A_1^2 = -I$, we have $d = \pm 1$ and we may choose it as 1. Hence

$$A = \begin{pmatrix} A_1 & 0 \\ 0 & 1 \end{pmatrix} \begin{pmatrix} I & 0 \\ \delta & 1 \end{pmatrix}.$$

But then $nAn^{-1} = \begin{pmatrix} A_1 & 0 \\ y(A_1 - I)+\delta & 1 \end{pmatrix}$ and since $A_1 - I$ is nonsingular, we may choose y so that $y(A_1-I)+\delta = 0$.

Corollary III.1.2. $J_S^4 = I$.

We will call an automorphism A of N satisfying the hypothesis of Lemma III.1.1 a C-R structure.

Definition. Let L be the 2m+1-dimensional Heisenberg algebra, let $\mathbb{T}$ be the complex torus $\mathbb{C}^m/D$ and let $L(\mathbb{T})$ be the Lie algebra of $N(\mathbb{T})$. The rational form of $L(\mathbb{T})$ determined by $H^+(\mathbb{C}^m/D, \mathbb{Z})$ will be called the D-rational form of $L(\mathbb{T})$. A Lie algebra morphism

$$\varphi^* : L(\mathbb{T}) \to L$$

is said to be of type H or an H-morphism with respect to D and J_T if

1. φ^* is a D-rational epimorphism, i.e., the kernel K of φ is a D-rational subspace;

2. $\varphi^*/\Lambda^1(V^*)$ is 1-1 and $\varphi^*/\Lambda^j(V^*) = 0$ for $j > 2$;

3. the kernel of φ, K, is J_T invariant.

We will now develop some elementary properties of H-morphisms.

P.1. An H-morphism induces a C-R structure on N.

Let J be the automorphism of L that J_T induces on $L(\mathbb{T})/K = L$. Then $\varphi(\Lambda^1(V^*))$ is J invariant, $\varphi|\Lambda^2(V^*) = \mathfrak{z}(L)$ and $J|\varphi(\Lambda^1(V^*)) = J_T(V^*)$. Hence J is a C-R structure on L and so on N.

P.2. An H-morphism determines a unique lifted subgroup of N.

Since φ is rational on $L(\mathbb{T})$, if we abuse notation by letting $\varphi : N(\mathbb{T}) \to N$ be the induced group surjection $\varphi(\Gamma(\mathbb{T})) = \Gamma_0$ is a discrete co-compact subgroup of N. But $\Gamma(\mathbb{T})$ has 2m generators, and so therefore does Γ_0. Hence Γ_0 is a lifted subgroup of N.

P.3. A choice of a generator of $\mathfrak{z}(\Gamma_0)$ determines an isomorphism between $N/\mathfrak{z}$ and $(N/\mathfrak{z})^*$.

The nondegenerate alternating form defining N determines a pairing of $N/\mathfrak{z}$ and $(N/\mathfrak{z})^*$ to $\mathfrak{z}$. Thus our assertion reduces to identifying $\mathfrak{z}$ and $\mathbb{R}$. Since $\mathfrak{z}$ and $\mathbb{R}$ are isomorphic, this is done by choosing a nonzero element of $\mathfrak{z}$ and sending it to $1 \in \mathbb{R}$.

It is classical that if $\varphi : S \to \mathbb{C}^m/D$ is the Jacobi imbedding, then $\varphi^* : L(\mathbb{C}^m/D) \to L(S) = L$ is an H-morphism. Indeed, one need only verify that φ is analytic and $\varphi^* H^1(\mathbb{C}^m/D, \mathbb{R}) \to H'(S)$ is an isomorphism. But these last facts are essentially the way that the Jacobi variety and imbedding are defined.

Let us now relate the concept of H-morphism to the more usual objects on complex tori: In order to simplify notation we will choose one of the generators of $\mathfrak{z}(\Gamma_0)$ and thus identify V and V^*.

Let $\varphi : L(T) \to L$ be an H-morphism relative to D and J_T. We can identify $\varphi|\Lambda^2(V) \to \mathfrak{z}(L)$ as a linear functional on $V \wedge V$. We denote $\varphi|\Lambda^2(V)$ by the symbol $\hat{A}$. The usual universal properties permit us to identify $\hat{A}$ with an alternating form A on V and thus a presentation of N given by

$$(v_1, t_1)(v_2, t_2) = (v_1+v_2, t_1+t_2+\hat{A}(v_1 \wedge v_2)).$$

We may ask: When does a codimension one subspace W of $\Lambda^2(V)$ determine an H-morphism? (Recall: Knowing the kernel is less than knowing the homomorphism. Many different homomorphisms have the same kernel. This will become clearer below.) The next few results will answer this question.

The following language is classical. The kernel of an H-morphism is called a polarization.

A codimension one subspace $W \subset \Lambda^2(V)$ provides us with a "line" in $\Lambda^2(V)^*$ consisting of those $\hat{A}$ with the property that the kernel of $\hat{A}$ equals W. A choice of any $\hat{A}$ determines a presentation of $V \times \mathbb{R}$, and this is a Heisenberg group if and only if $\hat{A}$ is nondegenerate.

Lemma III 1.3. Let the kernel of $\hat{A}$ in $\Lambda^2(V)$ be W. Then W is J_T invariant if and only if $\hat{A}_0 J_T = \hat{A}$.

Proof is obvious.

Lemma III 1.4. $\hat{A} \circ J_T = \hat{A}$ if and only if the bilinear form

$$(\xi, \eta) \to A(J\xi, \eta)$$

is symmetric.

Proof. Symmetry means that

$$A(J\xi, \eta) = A(J\eta, \xi), \qquad \text{all } \xi, \eta \in V.$$

The above is so if and only if

$$A(J(J\xi), \eta) = A(J\eta, J\xi).$$

But because A is an alternating form and $J^2 = -I$, we have the above is equivalent to

$$A(\xi, \eta) = A(J\xi, J\eta)$$

or

$$\hat{A}(\xi \wedge \eta) = \hat{A}(J\xi \wedge J\eta) = \hat{A}(J(\xi \wedge \eta)).$$

Lemma III 1.5. If D and V are as above, then W is D-rational if and only if $\hat{A}(D \wedge D)$ is discrete in $\underline{\underline{z}}$.

Proof. The statement that W is D-rational is equivalent to the image of $D \wedge D$ in $\Lambda^2(V)/W$ being discrete. But $\hat{A}$ induces an isomorphism $\Lambda^2(V)/W \to \underline{\underline{z}}$, under which the image of $(D \wedge D+W)/W$ is precisely the set $\{\hat{A}(d \wedge d') \mid d, d' \in D\}$.

In order to relate these results to more classical results let us recall the following definition.

Definition. Let D be a lattice and J a complex structure for the vector space V. By a Riemann form (in the weak sense of Swinnerton-Dyer) is meant a bilinear form

$$A : V \times V \to \underline{\underline{R}}$$

such that

1. A is alternating;
2. $A(D \times D)$ is a discrete subgroup of $\underline{\underline{R}}$;
3. The form $B(\xi, \eta) = A(J\xi, \eta)$ is symmetric.

The above results combine to yield

Theorem III.1.6. A Lie algebra morphism $\varphi : L(\mathbb{T}) \to L$ is an H-morphism with respect to D and J_T if and only if any form A of the polarization of φ is a Riemann form for D.

Definition. Let J be a C-R structure for L. Then J is called definite provided

$$[X, JX] = 0 \text{ if and only if } X \in \mathfrak{z}(L).$$

J operates on L. Let W equal the range of the linear operator (J-I). Let $X \in W$ and set $Y = -JX$. Since $J^2 = -I$, we see that X and Y are independent and span a 2-dimensional J invariant subspace U in W. Because J is semisimple on W, it follows that there is a basis $\{X_1, \ldots, X_n, Y_1, \ldots, Y_n\}$ for W over $\underline{\underline{R}}$ such that

$$J(X_k + iY_k) = i(X_k + iY_k), \qquad k = 1, \ldots, n,$$

$$J(X_k - iY_k) = -i(X_k - iY_k).$$

We shall call such a basis well aligned if, in addition, there is some $Z \in z(L)$ for which the following bracket relations hold:

$$[X_h, X_k] = [Y_h, Y_k] = 0$$

and

$$[X_h, Y_k] = \pm\delta_{hk} Z, \qquad 1 \leq h, k \leq n.$$

Let us introduce the notation

$$[X_h, Y_k] = \varepsilon_h Z.$$

<u>Lemma</u> III.1.7. Let B denote the bilinear form

$$B(X, X') = A(JX, X'), \qquad X, X' \in W.$$

Then B is definite precisely when there exists a well-aligned basis for which $\varepsilon_h = 1$ for all h or $\varepsilon_h = -1$ for all h.

<u>Proof.</u> Let $Z \in \underline{z}$ be such that

$$[X, X'] = A(X, X')Z, \qquad X, X' \in W.$$

The bracket relations defining well alignment then imply that the matrix of B with respect to a well-aligned basis is the diagonal matrix in which the entries corresponding to $B(X_h, X_h)$ and $B(Y_h, Y_h)$ are both ε_h. Thus B is definite if ε_h has constant sign. Conversely, if B is definite, and if there exists a well-aligned basis, then the ε_h will have constant sign. Thus, it remains to prove that when B is definite, there exists a well-aligned basis. The proof is by induction on n

Let X_1 be any nonzero vector in W and set $Y_1 = -JY_1$; because B is definite,

$$A(X_1, Y_1) = B(X_1, X_1) \neq 0.$$

Changing X_1 by a scalar factor if necessary, we have

$$[X_1, Y_1] = \pm Z.$$

Set

$$L_1 = \{U \in L \mid [U, X_1] = [U, Y_1] = 0\}.$$

Because A is J invariant, L_1 is invariant under J. Since B must be definite on L_1 we can now argue inductively that a well-aligned basis must exist.

Lemma III. 1. 7 has as one of its important consequences that we can use definite H-morphisms to identify V and V^*. By P. 3 all that is required is a selection of a generator of $\mathfrak{z}(\Gamma_0)$ or equivalently one of the components of $\mathfrak{z}(\Gamma_0)-\{0\}$. We will always choose the component in which Z points of $\varepsilon_h = 1$ and $-Z$ if $\varepsilon_h = -1$.

We will call a definite H-morphism positive if ε_h and negative otherwise. Clearly, if φ is a negative H-morphism and

$$A = \begin{pmatrix} 0 & I & 0 \\ I & 0 & 0 \\ 0 & 0 & -1 \end{pmatrix},$$ then $A \circ \varphi$ is a positive H-morphism.

The proceeding discussion proves that if $\mathbb{T}$ has a definite H-morphism, it satisfies the classical conditions of Frobenius for being an Abelian variety and conversely. This proves Theorem III.1.8. $\mathbb{T} = \mathbb{C}/D$ is an Abelian variety if and only if there exists a definite H-morphism $\varphi : L(\mathbb{T}) \to L$ with respect to D and J_T.

In the study of complex tori $\mathbb{C}^m/D$ one can regard a morphism of tori

$$\mathbb{C}^m/D \to \mathbb{C}^{m'}/D'$$

as a complex linear map $h : \mathbb{C}^m \to \mathbb{C}^{m'}$ such that $h(D) \subset D'$. The subspaces of $\mathbb{C}^m$ and $\mathbb{C}^{m'}$ that arise as kernels and images of such morphisms are rational with respect to D and D' respectively. The complete reducibility theorem of Poincaré gives a process for solving the following problem: Given an Abelian variety A and a subtorus A' find a subtorus A'' such that

$$A' \times A'' \to A$$

is an isogony. We will now use definite H-morphisms to give a new proof of this result.

Consider V a 2n-dimensional real vector space, J a complex structure for V and D a lattice subgroup of V. Let $N(A) = V \times \mathbb{R}$

be a presentation for the Heisenberg group with A alternating. We will say that

$$0 \to \mathbb{R} \to N(A) \xrightarrow{\pi} V \to 0$$

defines $N(A)$ as a Heisenberg cover for (V, J, D) provided

1. there is an automorphism $\hat{J}$ of N that induces J on V,
2. there is a lifted subgroup over D.

The following result is now clear.

Lemma III.1.9. A is a Riemann form with respect to (V, J, D) if and only if

$$0 \to \mathbb{R} \to N(A) \to V \to 0$$

is a Heisenberg cover of (V, J, D).

Suppose $N_A \xrightarrow{\pi} V$ is a Heisenberg cover of (V, J, D) and that W is a complex subspace of V, i.e., $JW = W$. Let $W^N = \pi^{-1}(W)$.

Lemma III.1.10. Let $N_A \xrightarrow{\pi} V$ be a definite Heisenberg cover. If D_1 is a sublattice of D, and the span of D_1 is V_1 and V_1 is J invariant, then D_1 has a lifted subgroup over it in N_A and (V_1, J_1, D_1) has

$$0 \to \mathbb{R} \to V_1^N \to V_1 \to 0$$

as a definite Heisenberg cover.

Proof. For the lifted group that covers D_1 we can take $\pi^{-1}(D_1) \cap \Gamma$. For $d_1 \in D_1$ let d_1' cover d_1 the fact that $[d_1', \hat{J}d_1] \neq 0$ assures us that $\pi^{-1}(D_1) \cap \Gamma$ is co-compact in V_1^N, and since it is a subgroup of Γ it is certainly discrete.

Lemma III.1.11. Let

$$0 \to z(N_A) \to N_A \to V \to 0$$

be a definite Heisenberg cover, then $W \subset V$ is D-rational if and only if $W^N \cap \Gamma$ is a discrete co-compact subgroup of W^N. In this case

$$1 \to z(N_A) \to W^N \to W \to 0$$

is a definite Heisenberg cover.

Proof. If $W \subset V$ is D-rational, then $W \cap D$ is a lattice subgroup of W and we can apply the lemma above.

Proposition III.1.12. Let $N_A \xrightarrow{\pi} V$ be a definite Heisenberg cover and let W be a complex subspace of V which is rational with respect to D. Further, let $C(W^N)$ be the centralizer of W^N in N_A (i.e., the set of all elements of N_A which commute with all the elements of W^N). Then $W^N \cap C(W^N) = z(N_A)$ and W^N and $C(W^N)$ generate N_A.

Proof. If $\alpha \in W^N$, because W is a complex subspace of V we have that $\hat{J}\alpha \in W^N$. Hence, for $\alpha \notin z(N_A)$ (the case $W = (0)$ is trivial), $[\hat{J}\alpha, \alpha] \neq 0$ implies that $\alpha \notin C(W^N)$. In the other direction, clearly $z(N_A) \subset W^N \cap C(W^N)$. We next observe that the statement that N_A is definite cover is equivalent to requiring that $H(\xi,\eta) = A(J\xi,\eta) + iA(\xi,\eta)$

is a definite Hermitian form. Further, the centralizer of W^N is precisely the inverse image under π of the H-orthogonal complement of W. These statements combine to prove our assertion.

Proposition III.1.13. Let (V, J, D) admit a definite Heisenberg cover $N_A \xrightarrow{\pi} V$ and let Γ be a lifting of D. Let W be a complex subspace of V which is D-rational. Then $C(W^N) \cap \Gamma$ is discrete and co-compact in $C(W^N)$.

Proof. The obvious fact that the centralizer of a rational algebraic group is rational and hence its intersection with Γ is co-compact.

We have included the above to indicate the appropriateness of the structure of H-morphisms to certain classical arguments. It is suggested that the reader compare the above to Theorem 34 in Swinnerton-Dyer [10].

III. 2. Nil-Theta Functions Associated with Negative H-morphisms of an Abelian Variety.

This section essentially reproduces material in Tolimieri [11]. It is justified by the need for a presentation consistent with the notation and normalizations that we have used in the rest of these notes. In order to make certain that we make choices consistent with our previous presentations, we will begin by considering a special case.

Begin with $\mathbb{C}^m$ and let D be the lattice of Gaussian integers in $\mathbb{C}^m$. Choose the presentation $\mathbb{C}^m \times \mathbb{R}$ for the Heisenberg group N(A) where $(c_1, \ldots, c_m, t_1)(c'_1, \ldots, c'_m, t_2) = (c_1+c'_1, \ldots, c_m+c'_m, t_1+t_2+\sum_i \operatorname{Im} c_i \bar{c}'_i)$. Let $c_i = a_i+b_i$, $i = 1, \ldots, m$, and $V^{2m} = (a_1, \ldots, a_m, b_1, \ldots, b_m)$. Then N(A) is given by the presentation $V^{2m} \times \mathbb{R}$, where A in the above coordinates has the matrix representation

$$A = \begin{pmatrix} 0 & \frac{1}{2}I_m \\ -\frac{1}{2}I_m & 0 \end{pmatrix} .$$

Let J be the automorphism of the complex structure on V^{2m}; then viewing J as acting on column vectors

$$J = \begin{pmatrix} 0 & -I_m \\ I_m & 0 \end{pmatrix} .$$

We will now verify that J is a positive C-R structure. Let X_i be the unit tangent vector in the increasing direction to the x_i coordinate curves and define Y_i, and T analogously. Then an elementary computation gives $[X_i, Y_i] = T$, $i = 1, \ldots, m$, and all other brackets are zero.

Now $J(X_i) = Y_i$, $J(Y_i) = -X_i$ and $J(T) = T$, $i = 1, \ldots, m$. Hence, $[X_i, JX_i] = [Y_i, JY_i] = T$ and J is a positive definite C-R structure.

Now let (N, Γ, J) be the image of a definite H-morphism h of an Abelian variety $\mathbb{C}^m/D$, where D is a lattice in $\mathbb{C}^m$. Then if J is not a positive C-R structure, we can choose an automorphism $\mathcal{A}$ of N such that $J(\Gamma) = \Gamma$ and $\mathcal{A}J\mathcal{A}^{-1}$ is positive. Thus we can always reduce our problem to the case of positive definite C-R structures.

The rest of this section will be devoted to generalizing the material in II. 1 to the case where $m > 1$. We will define and study the algebra of nil-theta functions of the triple (N, Γ, J), where J is a positive definite C-R structure. Here, however, we will not use complex analysis to establish the basic properties of existence and uniqueness of nil-theta functions, but we will establish these results using real analysis and distinguished subspace theory. This will require that we choose presentations of N that are suitable for using the Weil-Brezin map. This forces us to use general positive definite C-R structures.

[A word of warning is in order here. As I said, we are essentially following Tolimieri [11], but because he is working with negative definite C-R structures he had to use V_i eigenvectors where we will use V_{-1} eigenvectors to get the same operators and Tolimieri uses a different Weil-Brezin mapping from ours. This should enable the reader who so desires to pass back and forth from our presentation to Tolimieri's.]

Recall that $\mathfrak{A}^0$ denotes the group of automorphisms of N that preserve the orientation of the center of N.

Lemma III. 2. 1. Let K be a positive C-R structure. Then there exists $E \in \mathfrak{N}^0$ such that $EKE^{-1} = J$.

This is really another form of the well-aligned basis theorem.

Lemma III. 2. 2. Let all notation be as above and let

$$E = \begin{pmatrix} A & B & 0 \\ C & D & 0 \\ \delta_1 & \delta_2 & 1 \end{pmatrix}$$

be a C-R structure. Then E is positive definite if and only if B is symmetric negative definite.

Proof. A quick examination should convince the reader that we can really work modulo inner automorphisms or inside $Sp(2n, \mathbb{R})$. Let $E = \beta^{-1} J \beta$, where

$$\beta = \begin{pmatrix} M & N \\ R & S \end{pmatrix} .$$

Then $B = -{}^tSS - {}^tNN$ and so B is symmetric. Further, if $\underline{x} = (x_1, \ldots, x_m)$ we have $\underline{x} \cdot B\underline{x} = -(|Sx|^2 + |Nx|^2)$. Since ${}^tMS - {}^tRN = I$, we see that there is no nonzero $x \in \mathbb{R}^n$ such that $\rho\underline{x} = N\underline{x} = 0$. Hence, B is symmetric and negative definite.

We refer to Tolimieri [11] for a proof of the converse.

Lemma III. 2. 3. Let

$$E = \begin{pmatrix} A & B & 0 \\ C & D & 0 \\ \delta_1 & \delta_2 & 1 \end{pmatrix}$$

be a positive definite C-R structure. Then $B^{-1}A$ is symmetric.

<u>Proof.</u> Since α is a symplectic matrix $AB^t = BA^t$. Since B is symmetric and invertible, $B^{-1}AB^t = A^t$. Hence, $B^{-1}A = A^tB^{-1} = A^t(B^{-1})^t$ or $B^{-1}A$ is symmetric.

We will now use positive definite C-R structures to select nil-theta functions. Consider $L_{\mathbb{C}}$ the complexification of the Lie algebra of N and let E be a positive definite C-R structure. Let

$$L_{\mathbb{C}} = V_i(E) \oplus V_{-i}(E) \oplus \mathfrak{z}(L_{\mathbb{C}})$$

be the decomposition of the Lie algebra $L_{\mathbb{C}}$ into its eigenvector spaces with eigenvalues i, $-i$, 1 respectively, where $\mathfrak{z}(L_{\mathbb{C}})$ denotes the center of $L_{\mathbb{C}}$. Let

$$A(E) = \{F \in C^{\infty}(N) \,|\, V_{-i}(E)F = 0\}.$$

If

$$E = \begin{pmatrix} A & B & 0 \\ C & D & 0 \\ \gamma_1 & \gamma_2 & 1 \end{pmatrix} .$$

Then $V_{-i}(E)$ is found by solving the system of matrix equations

$$A\underline{x}+B\underline{y} = -i\underline{x}$$

$$C\underline{x}+D\underline{y} = -i\underline{y}$$

$$\gamma_1 \cdot \underline{x} + \gamma_2 \cdot \underline{y} + t = -it .$$

Since B^{-1} exists,

$$\underline{y} = -B^{-1}(iI-A)\underline{x}$$

$$t = \frac{-1+i}{2}(\gamma_1 \cdot \underline{x} + \gamma_2 \cdot \underline{y}).$$

We can now state and prove the main result of this section.

Theorem III. 2. 4. Let P be a principal subgroup of N and let E be a negative definite C-R structure on N. Then

$$\dim(A(E) \cap H_1(P)) = 1.$$

Proof. Since P is a principal subgroup, we may choose a presentation $(x_1, \ldots, x_m, y_1, \ldots, y_m, t)$ as above so that P is generated by $\gamma_i = (a_{i1}, \ldots, a_{i2m}, 0)$, $i = 1, \ldots, 2m$, where $a_{ij} = \delta_{ij}$, $i, j = 1, \ldots, 2m$. For $f \in \mathcal{S}(\mathbb{R}^m)$ the Weil-Brezin map $W : \mathcal{S}(\mathbb{R}^m) \to C^\infty(N) \cap H_1(P)$ is given by $W(f) = e^{2\pi i t} e^{\pi i \underline{x} \cdot \underline{y}} \sum_{\underline{\ell} \in \mathbb{Z}^m} f(\underline{x}+\underline{\ell}) e^{2\pi i \underline{\ell} \cdot \underline{y}}$. For $f \in \mathcal{S}(\mathbb{R}^m)$, we have, using the fact that W is an intertwining operator for U_1 and R_1, that

$$W\left(\frac{\partial f}{\partial s_j}\right) = X_j(Wf)$$

$$W(2\pi i s_j f) = Y_j(Wf)$$

$$W(2\pi i f) = T(Wf),$$

and we may talk of $W^{-1}(V)$, $V \in L_{\mathbb{C}}$.

Now, the discussion just prior to the statement of Theorem III. 2. 4 implies that for $V = \Sigma a_i X_i + \Sigma b_i Y_i + cT$ to be in $V_{-i}(E)$ we may choose

a_i arbitrarily, but then

$$\begin{pmatrix} b_1 \\ \vdots \\ b_m \end{pmatrix} = -B^{-1}(iI-A)\begin{pmatrix} a_1 \\ \vdots \\ a_m \end{pmatrix}$$

$$c = \frac{-1+i}{2}(\gamma_1 \cdot \underline{a} + \gamma_2 \cdot b).$$

Let $\underline{s}f = (s_1 f, \ldots, s_m f)$ and $\underline{v}_i^t = (\delta_{i1}, \ldots, \delta_{in})$, $i = 1, \ldots, m$, where δ_{ij} are the Kronecker delta. Then $V_{-i}(Wf) = 0$ is equivalent to the system of differential equations

$$\frac{\partial f}{\partial s_i} + (\underline{s}f) \cdot 2\pi i G\underline{v}_i + 2\pi i \frac{-1+i}{2}(\gamma_1 \cdot \underline{v}_i + \gamma_2 \cdot Gv_i)f = 0,$$

$i = 1, \ldots, m$, where $G = -B^{-1}(iI-A)$. Since the real part of $2\pi i B^{-1}(iI-A)$ is $-B^{-1}$ and this is positive definite, it is classical that this system of differential equations has a unique solution in $L^2(\mathbb{R}^m)$ up to scalar multiplication.

Theorem III.2.5. Let Γ be a discrete co-compact subgroup of N and let $\Theta_n(\Gamma) = H_n(\Gamma) \cap A(E)$, $n > 0$. Then

$$\dim \Theta_n(\Gamma) = \text{multiplicity of } H_n(\Gamma).$$

This theorem is an immediate corollary of Theorem III.2.4 and the facts about distinguished subspaces in $H_n(\Gamma)$.

III. 3. The Relation Between Nil-Theta and Classical Theta Functions.

The formulas and results in this section are all in Tolimieri [11], and we would not have included this section except for the fact that there is a construction in [11] that we feel that can be clarified using the concept of presentation of the Heisenberg group. Accordingly, we will develop just enough material to obtain this additional insight into the results in [11].

We will begin by reviewing the elementary classical theory of Abelian varieties. For a lattice $D \subset \mathbb{C}^m$ we will call a nondegenerate R-bilinear form $A : \mathbb{C}^m \times \mathbb{C}^m \to \mathbb{R}$ a nondegenerate Riemann form if

a) A is alternating,

b) $A(D \times D) \subset \mathbb{R}$ is a discrete subgroup,

c) $(\underline{z}_1, \underline{z}_2) \to A(i\underline{z}_1, \underline{z}_2)$, $\underline{z}_i \in \mathbb{C}^m$, $i = 1, 2$, is a positive definite $\mathbb{R}$ form. [Note: The automorphism J corresponding to multiplication would be negative definite by the definition we used for CR structures.]

Clearly, not every lattice $D \subset \mathbb{C}^m$, $m > 1$, admits a nondegenerate Riemann form. A nondegenerate Riemann form is easily seen to determine a unique definite H-morphism and visa-versa by the material in III. 1. By I. 1, A determines a presentation of $\mathbb{C}^m \times \mathbb{R}$ or $V^{2m} \times \mathbb{R}$. As usual, we will view (V^{2m}, J_{2n}) as $\mathbb{C}^m$.

Lemma III. 3. 1. Let A be a $2m \times 2m$ nondegenerate real form on V^{2m} that determines a nondegenerate Riemann form. Then there exists a real nonsingular matrix C such that

$$A = C^t J_{2n} C$$

$$J_{2n} C = C J_{2n}.$$

Remark. Hence C determines a complex linear transformation on $\mathbb{C}^m$ that takes A to J_{2n} and D to $C(D)$ and leaves the automorphism determining the C-R structure, J_{2n}, fixed.

Proof. Since $A^t = -A$, AJ_{2n}^t is symmetric and $J_{2n}^t = -J_{2n}$, we have

$$AJ_{2n} = -AJ_{2n}^t = -J_{2n}A^t = J_{2n}A.$$

Hence, $J_{2n}^t A = AJ_{2n}^t$ is positive definite. It follows that if

$$A = \begin{pmatrix} X & Y \\ -Y & X \end{pmatrix}$$

and $C_1 = C_1' + iC_1''$, we have

$$Y + iX = \overline{C}_1^t C_1 .$$

Then letting

$$C = \begin{pmatrix} C_1' & -C_1'' \\ C_1'' & C_1' \end{pmatrix}$$

this proves the lemma.

This lemma says that the classical theory works with the negative of the basic presentation and negative C-R structures. The automorphism

$$\begin{pmatrix} 0 & I & 0 \\ I & 0 & 0 \\ 0 & 0 & -1 \end{pmatrix}$$

of N exchanges the classical hypothesis with the one we have chosen to work with. Accordingly, if we study the basic presentation with J_{2n} as C-R structure, this is equivalent to the classical study.

Accordingly, we will study the special case when

$V^{2m} = (x_1, \ldots, x_m, y_1, \ldots, y_m)$,

$$A = \begin{pmatrix} 0 & \frac{1}{2}I \\ -\frac{1}{2}I & 0 \end{pmatrix}$$

$$J_{2m} = \begin{pmatrix} 0 & -I & 0 \\ I & 0 & 0 \\ 0 & 0 & 1 \end{pmatrix}$$

and Γ is the principal subgroup generated by

$$\gamma_i = (a_{i1} \ldots a_{i2m}), \qquad \text{where } a_{ij} = \delta_{ij},\ i, j = 1, \ldots, m,$$

and δ_{ij} is the Kronecker δ. We begin by generalizing Theorem II. 1. 3.

Let $\Theta_n(\Gamma, J_{2m})$ be the nil-theta function in $H_n(\Gamma)$, $n > 0$. Let $M_n(\underline{x}, \underline{y}, t)$ be defined by

$$M_n = M_n(\underline{x}, \underline{y}, t) = e(-2\pi i n t)e(-\pi i \underline{x} \cdot \underline{y})e(\pi n \underline{y} \cdot \underline{y}).$$

Let $\theta_d = \Theta(J_{2n}) \cap C_d^\infty(N)$ where $C_d^\infty(N)$ consists of all $F \in C^\infty(N)$ of the form

$$F(\underline{x}, \underline{y}, t) = e^{2\pi i t d} F(\xi, 0), \qquad d \in \mathbb{R},\ d > 0,$$

and $\Theta(J_{2n})$ denotes the nil-theta functions determined by the C-R structure J_{2n}.

Theorem III. 3. 2. Let $E(\mathbb{C}^m)$ denote the entire functions on $\mathbb{C}^m$. Then the mapping $F(\underline{x}, \underline{y}, t) \to M_d(\underline{x}, \underline{y}, t)F(\underline{x}, \underline{y}, t)$ defines an isomorphism from θ_d onto $E(\mathbb{C}^m)$.

Outline of Proof. Since $(\underline{X}+i\underline{Y})F = 0$ we have

$$\pi d(\underline{x}+i\underline{y})F_1 + (\frac{\partial}{\partial x}+i\frac{\partial}{\partial y})F_1 = 0,$$

where $F_1(\underline{x}, \underline{y}) = F(\underline{x}, \underline{y}, 0)$. Computing yields that $M_d F = e(\pi d\underline{y}\cdot \underline{y})e(-\pi i d\underline{x}\cdot \underline{y})F_1$ is a solution of the Cauchy-Riemann equation. The converse follows by reversing the steps.

Corollary. $M_n\Theta_n(\Gamma, J_{2m}) \subset E(\mathbb{C}^m)$.

We will now see that the Γ periodicity of $F \in \Theta_n(\Gamma, J_{2m})$ implies that $M_n F$ satisfies certain functional equations. As in II. 1 we have that $M_n\Theta_n(\Gamma, J_{2m})$ consists of all entire functions on $\mathbb{C}^m$ satisfying the functional equations

$$\theta(\underline{\xi}+\underline{\eta}) = \theta(\xi)e(-2\pi i n L(\underline{\xi}, \underline{\eta}))e(-\pi i n L(\underline{y}, \underline{y})),$$

where $\underline{\xi} = \underline{x}+i\underline{y} \in \mathbb{C}^m$ and $\underline{\eta} \in D$ and $L(\underline{\xi}, \underline{\eta}) = \underline{s}\cdot \underline{x} + i\underline{s}\cdot \underline{y}$ where $\underline{\eta} = \underline{r}+i\underline{s}$.

As before, if we vary Γ amongst all Γ such that $\pi(\Gamma) = D$ we have the action of inner automorphisms on N and the functional equations change by a character. The essentially new element is the effect of variations of presentation, that we will now discuss.

Let G be a symmetric bilinear form and consider the bilinear form $G+A$ and let $N(G)$ denote the presentation determined by $G+A$. Then the isomorphism

$$\psi : N \to N(G)$$

is given by $(\underline{x}, \underline{y}, t) \to (\underline{x}, \underline{y}, t+\frac{1}{2}G((\underline{x}, \underline{y}), (\underline{x}, \underline{y})))$.

Consider $F : \psi(\Gamma)\backslash N(G) \to \mathbb{C}$. Then, since ψ is an isomorphism, we have

$$\begin{array}{ccc} N & \xrightarrow{\psi} & N(G) \\ & {\scriptstyle f}\searrow & \big| {\scriptstyle F} \\ & & \mathbb{C} \end{array}$$

so that $F(\xi, t) = f(\psi^{-1}(\xi, t))$ where $\xi = (\underline{x}, \underline{y})$. But

$$\psi^{-1}(\xi, t) = (\xi, t - \tfrac{1}{2}G(\xi, \xi)).$$

If $f \in \Theta_n(\Gamma)$ or $H_n(\Gamma)$, we have

$$F(\xi, t) = \exp(-2\pi i n \tfrac{1}{2}G(\xi, \xi))f(\xi, t).$$

Now for $f \in \Theta_n(\Gamma)$, $M_n f$ is entire. We may ask when is $M_n F$ entire? Clearly, $M_n F$ is entire if and only if G is $\mathbb{C}$ linear. Assume that G is $\mathbb{C}$ linear. Then

$$F(\xi+\eta, 0) = \exp(-2\pi i n \tfrac{1}{2}G(\xi+\eta, \xi+\eta))f(\xi+\eta, 0)$$

or

$$F(\xi+\eta, 0) = \exp(-2\pi i n \tfrac{1}{2}(2G(\xi,\eta) + G(\eta,\eta)))F(\xi, 0).$$

It follows that $M_n F$ satisfies the following functional equation:

$$(M_n F)(\xi+\eta) = e(-2\pi i n(L(\xi,\eta)+G(\xi,\eta)))e(-\pi i n(L(\eta,\eta)+G(\eta,\eta))MF(\xi),$$

where $\eta \in D$ and L is as above.

We now will close this section, leaving it for the interested reader to follow up by reading Tolimieri [11].

BIBLIOGRAPHY

[1] L. Auslander, An exposition of the structure of solvmanifolds, Part I. Algebraic Theory. *Bull. A. M. S.*, 79 (1973), 227-261.

[2] L. Auslander and J. Brezin, Translation invariant subspaces in L^2 of a compact nilmanifold I, *Inventiones Math.*, 20 (1973), 1-14.

[3] L. Auslander and R. Tolimieri (assisted by H. E. Rauch), *Abelian Harmonic Analysis, Theta Functions and Function Algebras on a Nilmanifold*, Lecture Notes in Math., 436, Springer-Verlag, 1975.

[4] L. Auslander and R. Tolimieri, Algebraic structures for $\oplus \Sigma_{n \geq 1} L^2(\mathbb{Z}/n)$ compatible with the finite Fourier transform (preprint).

[5] J. Brezin, Harmonic analysis on nilmanifolds, *Trans. Amer. Math. Soc.*, 150 (1970), 611-618.

[6] J. Brezin, Function theory on metabelian solvmanifolds, *J. of Functional Analysis*, 10 (1972), 33-35.

[7] C. Kosniowski, *Transformation Groups*, Cambridge Univ. Press, 1977.

[8] S. Lang, *Algebra*, Addison-Wesley, 1965.

[9] S. Lang, *Introduction to Algebraic and Abelian Functions*, Addison-Wesley, 1972.

[10] H. P. F. Swinnerton-Dyer, *Analytic Theory of Abelian Varieties*, Cambridge University Press, 1974.

[11] R. Tolimieri, Heisenberg manifolds and theta functions, to appear Trans. A. M. S.

[12] A. Weil, L'Integration dans les Groupes Topologique et ses Applications, Hermann, Paris, 1953.

[13] A. Weil, Sur certaines groupes d'operateurs unitaires, Acta. Math., 111 (1964), 143-211.

[14] A. Weil, Variétés Kählériennes, Actualités Scientifiques et Industrielles, 1269, Hermann, 1958.